PRAISE FOR T

"There are definitely keys to selecting and working with a digital agency, and Drew has laid it out for you. This is a straightforward, honest, and easy-to-understand guide that focuses on a partnership from a client's perspective but with the agency's insights. After you read it, keep it close by. Because you'll find it's not just a great primer, it's a valuable resource."

—Carl Smith, Bureau of Digital

"Thorough, approachable advice for a process that can otherwise seem daunting. This clear guide will answer all of your questions about paying someone to build your website—including the ones you didn't know to ask."

—Mark Jaquith, Lead WordPress Developer

THE BUYER'S GUIDE TO WEBSITES

What You Need to Know to Get
the Website of Your Dreams

DREW BARTON

Copyright © 2017 Southern Web, LLC

All rights reserved.

No part of this book may be reproduced, or stored in a retrieval system, or transmitted in any form or by any means, electronic, mechanical, photocopying, recording, or otherwise, without express written permission of the publisher.

Published by Southern Web, LLC, Atlanta, Georgia
www.southernweb.com

Edited and Designed by Girl Friday Productions
www.girlfridayproductions.com

Editorial: Emilie Sandoz-Voyer
Interior and Cover Design: Rachel Marek
Image Credits: cover © GLYPHstock/Shutterstock

ISBN (paperback): 978-0-9993224-0-6
e-ISBN: 978-0-9993224-1-3

First Edition

Printed in the United States of America

TABLE OF CONTENTS

PART ONE: KNOWLEDGE IS POWER

1. What You Don't Know Can Hurt You 3
2. The Keys to Success 9
3. Choosing a Digital Agency: Questions to Ask . . 21
4. Choosing a Digital Agency: Red Flags to Avoid . 31
5. The Most Important Question 37

PART TWO: THE LEGAL STUFF

6. The Fine Print: Know What's in Your Contract . 43
7. Understanding Statements of Work and Service Level Agreements 53
8. Avoiding Scope Creep 61
9. When Things Go Wrong 71

PART THREE: BUILDING A WEB PRESENCE THAT WORKS

10. Budgeting for Your Website 81
11. Building and Developing Your Website 89
12. The Ins and Outs of Digital Marketing 99
13. Email Marketing 113

EPILOGUE: THE TAKEAWAYS 123

APPENDIX	. 127
GLOSSARY	. 131
ACKNOWLEDGMENTS	. 143
ABOUT THE AUTHOR	. 145

PART ONE: KNOWLEDGE IS POWER

CHAPTER 1

What You Don't Know Can Hurt You

For many of us, it starts with a dream—namely, the dream of having our own business.

In Janie's case, her dream was to have her own line of skin-care products. She'd spent some time experimenting and researching and had managed to develop a few creams and lotions. After selling to her friends and at a few trade shows, she was convinced it was time to take the leap. Rather than buying or renting a storefront, Janie decided to try to reach a global market by taking her products online.

The only problem was that Janie knew very little about the Internet and how it works. She didn't know what goes into making a website, and she knew even less about ecommerce. She could do a Google search and check her email, but that was about it. She began looking for someone to help her, and landed upon a friend of a friend who told her he could create her website on the cheap. Excited and a little desperate to get going, she hired him and put down a hefty deposit.

From that point, things began to spiral.

The web designer she hired promised to have the website live within a few weeks, but weeks turned into months with no progress. Every time she called to check, he said he was running a little behind, but not to worry, it would be up soon. He never ran the designs by her, nor did she think to ask.

When her website finally launched, it looked shoddy—nothing like what they'd discussed in their first meeting. Half the buttons wouldn't work. The shopping cart was broken. It looked unprofessional, and functioned even worse. It was embarrassing.

Fuming, Janie called the designer, but his phone had been disconnected. He had gone out of business, and she didn't know how to reach him. She now had a broken website for everyone to see with her name, business name, logo, and product line on public display, and she didn't even know how to access it to take it down. Worst of all, because she'd relied on a verbal agreement instead of a written contract, it would be difficult to sue the designer, even when or if she found him.

Janie's story sounds a little extreme, but it's more common than you think. I wish I could tell you this is the worst-case scenario, but it's not. Like Janie, countless people have found themselves in similarly unpleasant situations, because they just didn't know what they were walking into.

You know that saying, "What you don't know can't hurt you"? It's not true, especially in the realm of the Internet. What you don't know *can* hurt you. Badly.

THE NEW WILD WEST

Everyone knows there is plenty of bad stuff on the Internet. In many corners, it's just not a pleasant place to be, to put it mildly. There's hate speech. There's porn. There are trolls and subtweeters. There are hackers, phishers, and thieves. In addition to these threats, there are plenty of unqualified or dishonest people in the web-development community that are just waiting for their chance to prey upon the unsuspecting.

That said, the Internet isn't inherently evil. It's just not regulated by a governing body that sets quality-assurance mandates. The online world is new and didn't even exist just a couple of decades ago. Internet use continues to expand at a remarkable rate, yet protections of the medium and its users remain few in number.

The thing about this new frontier is that participation isn't optional for anyone who is serious about doing business in this day and age. If you're in business, it's assumed you'll have a website, which means you have to learn how to survive on the digital frontier without becoming a victim of bandits or outlaws.

If you think about it, pretty much every industry struggled with incompetence and predatory practices before there were regulations to guide it in an equitable direction. Before building codes were established, people were constructing buildings that often caught fire. Before the advent of medical associations, physicians practiced without common guidelines. At the same time, there were also *good* doctors who treated the sick using the best information available, and there were *good* carpenters who built houses that are still standing centuries later.

By the same token, not every digital agency is going to screw you over or deliver substandard work. There are very good agencies out there, with people who have taken the time to learn coding, graphic design, and digital best practices. These are the type of people who make it their priority to stay up with the latest trends and protect their clients. They are the people you can trust to stand behind their work and offer a solid, honest contract to their customers.

So along with the bad, there is good. Despite the lack of organizations that regulate the quality of a digital agency, you can find good agencies in this Wild West that will build you a great website. The key is to know how to find them, what questions to ask, and how to protect your interests. That's where this book comes in.

KNOWLEDGE IS POWER

The tragedy in Janie's story wasn't just what happened with her website. It was also that she was effectively powerless to stop it. Janie's biggest vulnerability—and what made it easy for the web designer to take advantage of her—was her lack of knowledge. Knowledge is power, and conversely, lack of knowledge is a weakness. Janie didn't need to be versed in computer coding, but had she at least known what questions to ask the web designer, she could have seen from the very start that this was not a guy with whom she should do business. Had she known to ask for a contract, and had she actually known what should be *in* the contract, she would have at least had some legal recourse when things went wrong.

Until the digital Wild West is tamed, you need to protect yourself. The way to do that is to educate yourself on the basics of how the web-development and design process works. You don't need to know what the developers know, but you *do* need to know how to work with a developer, and most importantly, how to tell a good one from a bad one. In order to make the most of your investment, you also need to know a little bit about how your new website should work, how to help your customers find it, and how to leverage it to increase your traffic and sales.

I wrote this book to empower business owners with the knowledge required to make smart web-development project decisions. In the chapters ahead, we're going to cover all the bases to ensure that you not only survive this new frontier, but thrive in it.

> **Knowledge is power, and conversely, lack of knowledge is a weakness.**

Rest assured, I'm not going to burden you here with all the technical stuff that goes into website development. After all, your job isn't to learn how to design websites. Your job is to run your business, and you should be able to do that without taking a crash course on Internet technology, digital marketing, and design principles. Besides, if you are interested in that sort of thing, there are plenty of other great books on the subject.

I *will*, however, tell you what you need to know, and what you need to *do*, in order to protect your interests when developing a web presence. You won't necessarily be an Internet guru when we're done, but you'll definitely know enough about the process to prevent yourself from becoming a victim.

To keep things simple, I've divided the material into three sections.

In Part One, we'll outline the digital agency's role in developing and designing your website, cover the questions to ask when choosing an agency, and talk about how you can most effectively work with an agency to achieve the best possible results.

In Part Two, we'll discuss the importance of protecting yourself legally when working with an agency. By demystifying the key elements of your contract—such as the Statement of Work and Service Level Agreement—I'll show you how a contract protects you when things go wrong.

In Part Three, we get into the nuts and bolts of planning your website, from drawing up and sticking to a budget to leveraging your web presence to grow your business through digital marketing. We'll finish with some closing thoughts and provide some helpful additional resources for you to look at moving forward.

The Internet is brimming with scammers, but when you are armed with the right knowledge, you'll discover that it's also home to ripe opportunities for your business to grow. So many businesses are already achieving great success online—it's now more important than ever to join them.

CHAPTER 2

The Keys to Success

Unless you plan to utilize one of the options available for creating a website on your own (more on that in Chapter 10), chances are you'll need to hire a professional to help you build your site. Many people and companies offer these types of services, and their role in your process can vary from simple consulting to full design, build, and marketing services.

To make things even more complicated, you'll sometimes hear these professionals described as *web designers* and other times as *web developers*. These terms are often used interchangeably, but they can actually mean different things. A web designer typically handles front-end development tasks to create what you visually interact with, while a web developer often handles the back-end construction, which is less about making the site pretty and more about ensuring it's functional. However, some developers, called *full-stack developers*, are well versed in both front- and back-end development, and some designers are familiar with back-end programming languages. In an effort to keep things simple, we'll be using the blanket term *digital agencies* to refer to third-party individuals or companies who play a role in helping you set up your website.

When hiring a digital agency, keep in mind that you are hiring an agent for your organization that will serve as a guardian of your brand, your online presence, and your online marketing. It's important to know what your responsibilities as a client are in that relationship, as well as what the responsibilities of the digital agency are. This relationship needs to be well defined during the proposal phase, long before you sign an agreement. So let's break it down.

Here are some questions that you, the digital agency's client, should be able to answer clearly early on in the process:

- What are you hiring the agency to do for you?
- What are the deliverables? Who provides them?
- When are the deliverables due?
- How much is the budget for additional expenses like stock photography, online advertising, and print marketing?
- What will the timeline of the project roughly look like?
- Who will give final approval on the various milestones of the project?
- What happens when there's a disagreement among the project's decision-makers?
- Will the digital agency be involved in discussions about design? Will the project manager be part of the decision-making process in order to understand the rationale behind the design decision?

During this process, you'll also need to clarify the roles of individuals working on both the client and agency side. This is especially important if you are part of a larger organization with multiple decision-makers. Too many cooks in the kitchen can needlessly complicate or hinder the project.

When interviewing and selecting a digital agency, you want to look at their experience, their overall process, and how well they have performed on other projects. Their process or workflow needs to match the way your organization makes decisions. If you have an organization that is very fluid—for example, you often make decisions quickly and then revise those decisions along the way—you might need to allow for additional time for revisions and feedback. You should also expect to pay the agency more for the extra time they would be putting in to accommodate your fluid workflow.

COMMUNICATE, COMMUNICATE, COMMUNICATE

Once you've chosen a particular digital agency and you've outlined all respective roles, one of the most important things you can do is to establish clear channels of communication between your organization and the agency. What is the best way for the agency to communicate with you about your project? Is it strictly by email or by phone? Will the agency only communicate at certain milestones in the project, or will there be daily or weekly check-ins with your team? Make sure these parameters are clear between both parties before entering into an agreement, as this can prevent a lot of confusion later on.

Communicate specific needs and wants

Website development and design are not simply about coding; they are creative processes. When you hire a digital agency to build a website for you, you're working with creative professionals who have ideas that might not necessarily match what you would do if you had the ability to build the site yourself. In many cases, it helps to keep an open mind about the site they design for you. That said, if you have certain firm boundaries, special needs, or deal breakers—things your website *must* or *must not* have—communicate these needs to the agency clearly and early in the process, if not before signing the agreement. If you know that your firm has certain sensitive preferences—for example, your logo should never be put on a black background because the boss hates it—that fact needs to be communicated to the agency in the early phases of the project. Don't allow the agency to make the mistakes that you know ahead of time are going to be issues. Otherwise, it can diminish the trust between you and the agency.

Sometimes these special needs aren't just about personal preference. Sometimes they can be uncovered in more unusual ways. Let me give you an example.

For many years, we worked with a marketing consultant who specialized in assisting law firms develop their marketing collateral, such as brochures, email marketing, and their website. On five separate engagements, the marketing coordinator brought on my agency to assist them with their digital marketing needs. Each time he engaged us, it was the same recipe: "I want the colors of the website to be blue, gray, and white." Every law-firm site we developed for him was always blue, gray, and white. Over and over again. Ad blue-gray-and-white nauseum.

The marketing consultant called us for a sixth engagement. Another law firm. Again, he asked for blue, gray, and white. By that point, we were honestly exhausted of doing blue, gray, and white sites for no good specified reason. Our designers half-heartedly joked that we were going to start a division called Blue Websites"R"Us.

In the absence of any rationale for why every website needed to be blue, gray, and white, our creative team made a decision to try a different color scheme—this time a bright crimson red—and pitch it to him. We sent over the design concept, and five minutes later, the marketing consultant called, irate.

"Why are you yelling?" I asked.

"This is not what we asked for. This is not what we like," he said. "You've built many websites for me and my clients, but this one is atrocious and I could never show it to my client. It's unsightly."

I asked a simple question: "Why does every single website we design for you, regardless of the client, need to be blue, gray, and white?"

He replied, "I'm color-blind." He had been asking for blue, gray, and white websites because those were the only colors he could differentiate.

In the history of humankind, countless projects have fallen apart for one of two reasons: communication or money.

Bear in mind that by this point we had a long-standing relationship with this consultant, and this issue had the potential to ruin the trust we'd built, all because of a key detail we couldn't possibly have known without being told. The website decision-maker's color blindness was a factor that affected the creative direction of the project. Had we been told this in the initial stages of working with him, our creative team would have been better equipped to either stay within his preferred color palette or discuss creative alternatives within that context. Instead, I basically walked into a minefield by delivering a project with a color scheme that was abhorrent to his sensibilities. To avoid a scenario like this one, clearly communicate any special considerations or nonnegotiables with your digital agency. Preferably, you would do this in your first few discussions—long before the design process begins.

Also, if there's a deadline for your website to go live—for example, in advance of a company event, such as a trade show or a year-end meeting—make sure the digital agency is aware of the deadline date and the reason for it before signing the agreement. Springing an unreasonable deadline following the acceptance of the agreement can deteriorate the relationship between you and the agency.

If you prefer to communicate by phone instead of email, it's important that you tell the agency so they communicate with you in your preferred medium. The same holds true with your website. If you prefer to receive sales leads via email or a contact form, it may be best to hide or make it difficult to find your phone number.

For the majority of reputable digital agencies, most requests are usually fine as long as they are communicated in advance, whether they are "Please avoid any mentions of a discount," or "The boss really hates orange." It doesn't matter what it is. The important thing is to *communicate it*.

Why are we harping on this? Because in the history of humankind, countless projects have fallen apart for one of two reasons: communication or money. We'll talk more about money in the budgeting chapter, but for now, it's vital to understand that proper communication—or the lack thereof—can make or break your relationship with your digital agency. This is especially true when you have multiple decision-makers on your project. If you don't communicate key information with the agency and they make a significant error as a result, it can damage the trust your team has in the agency, and the agency ends up receiving the blame because of stipulations they couldn't possibly have known about.

DIGITAL AGENCY AS REPRESENTATIVE

Here's another way to look at it: think of your digital agency as your attorney. Just as a lawyer represents your interests in business transactions and in court, your digital agency represents your interests on the Internet. The agency exists to interpret your company's image and vision in the form of a web presence. This means that even as you trust the digital agency to perform the technical and creative tasks you often can't do yourself, it's up to you to make sure the agency has all the important information needed to represent your interests correctly and to act on your behalf. This is where communication matters most.

Swim in Your Lane

One of the biggest issues that can cause a communication breakdown between you and the agency is when defined roles become muddled.

Imagine a race where forty swimmers are all lined up at one end of a massive swimming pool. Due to the sheer number, each racer is elbow to elbow, shoulder to shoulder, hip to hip. It's a tight squeeze.

Suddenly, the signal is given to start the race. At first, half the swimmers jump in, closely followed by more. Arms are thrashing, legs are kicking, water is splashing. It's an amazing sight to see.

Moments later nearly everyone got to the other side of the pool. Most of the swimmers have been kicked, elbowed, or scratched. None of them could possibly set record times because of the wake caused by the other swimmers and the jabs from their strokes.

After a few of these haphazard aquatic debacles, the race officials realize that it might be better to give each swimmer their own lane. This would allow swimmers to finish their race without interfering with each other.

The result is a smooth swim. Swimmers arrive unfettered at the other end of the pool. They aren't elbowed, nudged, or jabbed. They set time and distance records, both personal and worldwide. Swimming races becomes a pleasure.

All of this happens because of swim lanes. Of course, in business settings, the notion of swim lanes isn't always a thing.

Every creative professional has been in a meeting where we've heard one of these gems:

- "I showed your work to my daughter-in-law who went to art school, and she gave me a few notes."
- "I don't know much about this tech stuff, but . . ."
- "It shouldn't take you very much time . . ."

Situations like these demonstrate the dire need for swim lanes for you and the agency. When respectful boundaries and expectations are set and prioritized throughout the process, you are explicitly stating the division of labor and the roles of each person in the project. Your proposal or Statement of Work should explicitly state what is expected of both parties, such as when you must be available to provide feedback, as well as when the agency will provide the deliverables.

You'll also want to cover with your digital agency what happens when one party fails to deliver—whether that be feedback or work. Some agencies will have clauses in their agreements to pause the work, and some will have a penalty charge for missed meetings or deadlines.

Once the "swim lanes" are designated, it's now time to let the professional do their job. If you hired the digital agency because they do fantastic work, then going back and forth on where every article, pixel, and graphic should be placed probably defeats the purpose of hiring the agency. Once you've decided to entrust your project to the agency, and once the roles of client and agency are defined, that trust should stay with the agency unless there is sufficient reason not to trust them anymore.

Continuing with our swim-lane analogy, I'd like to mention one more thing required to ensure a successful working relationship with a digital agency. Once you've entered your swim lane, you need to *swim*. That means keeping pace alongside the agency, coordinating your efforts with theirs, and following the timetables or benchmarks the two of you have set together. If you don't do your part, the process can and will get hung up, which can lead to misunderstandings. Think of it as a relay race. If you haven't met a deadline for something you needed to provide the agency, you can't expect the agency to meet their own deadline. In that scenario, they are essentially standing at the other end of the pool, waiting for you to swim up to them and tap them into the next phase of the effort.

To illustrate the importance of this point and how it can affect the relationship, let me share a quick story. We once had a client who needed a website, and in determining our respective roles, the client decided that they would provide their own copy instead of paying a copywriter. In order for us to move forward on our end, the client agreed to send thirty pages of copy on a given date. For whatever reason, they didn't send us the copy, and as a result, we missed our timetable for the next phase of the project.

Soon after, I received an email from the president of the organization saying that they were not going to pay the next installment at the determined due date, because we had not completed a milestone within the established time frame—despite the fact that we had not received the copy that they had promised.

In our contract, payments were set by date—not by milestones determined by asset delivery. In other words, the client had agreed to pay installments by certain calendar dates, regardless of how the project was progressing. So while the client was arguing he shouldn't have to pay because we were behind schedule, from our end of the agreement, the payments were due when they were due. To make things even hairier, our agency has a policy not to move forward on projects that are in default. This put us at even more of a standstill, and the relationship became even more strained as a result. All the client needed to do to break the logjam was "swim up" to us and hand off those thirty pages of content. The holdup wasn't due to slacking off on our part; rather, the client had neglected to swim in their lane.

WHAT IF THE AGENCY IS BAD?

Throughout this chapter, we've been discussing the best ways for you as a client to deal with a digital agency. Of course, we've also been operating on the general assumption that the agency you've hired is reasonably competent and honest, and truly wants to provide the services you need.

However, as we said in the previous chapter, the Internet is the new Wild West. And while there are plenty of agencies who dedicate themselves to going the extra mile for their clients, there are many others out there who are neither competent nor honest.

There are countless tales of business owners who hire sub-par agencies and find themselves with badly designed, broken websites—or no website at all—with little or no recourse for getting their money back. Never mind the whole analogy of "swimming in your lane." With poorly managed agencies, you're not likely to have any lanes at all. What's most worrying is the fact that unless you have a keen understanding of Internet technology and terminology going into the process, the sales pitch of a bad agency can sound exactly the same as that of a good one!

◆ ◆ ◆

So how can you protect your interests? How can you tell the difference between an aboveboard agency and a shady one? That's the key question we'll be answering in the next two chapters—starting with questions you need to ask when vetting an agency, followed by some major red flags to avoid.

CHAPTER 3

Choosing a Digital Agency: Questions to Ask

Does the following scenario sound familiar?

While wandering the aisles of a grocery store, you observe a little girl talking to her mother. The girl, who can't be older than three, is animated and full of energy. The mother looks tired, even a little exasperated. You don't mean to eavesdrop, but you do it anyway. The conversation sounds something like this:

"Mommy? Mommy, what's that?"

"That's a can opener, dear."

"What does it do?"

"You use it to open cans."

"Why?"

"Because the cans are sealed, and you need something to open them."

"Why?"

"It keeps the food fresh."

Satisfied for now, the little girl runs to a nearby display.

"What kind of fruit is this, Mommy?"
"That's a cantaloupe."
"Does it taste good?"
"Yes."
"It looks yucky on the outside. Why does it look like that?"
"I don't know, honey. That's just what it looks like."
"Can I have these cookies?"
"No."
"Why?"
"They're bad for your teeth."
"Why?"
"The sugar makes your teeth rot."
"Why?"

By now, you're probably nodding knowingly. Nearly everyone has witnessed an exchange like this at some point, whether it's in the grocery store, a waiting room, or a line at an amusement park. If you've ever cared for small children, you've probably experienced this for yourself more than a few times. Young children are full of questions—so many, in fact, that sometimes you might wish for just a *few* quiet moments without all the questions.

And yet, questions are critical to the child's growth. One of the key ways we learn about the world around us is by asking questions. Lots of them. This approach doesn't change as we get older, even though we hopefully learn to pace the questions a bit. As soon as we stop asking questions, we stop learning, and we stop growing.

So, enough philosophy. You get the idea.

The point is that when you're vetting a digital agency to build and/or host your website, you're going to learn about them the same way the three-year-old child learns—by asking questions.

Remember in the previous chapter when we talked about thinking of the digital agency you hire as a representative of your company? The best way to know whether you can trust any person or any company to represent your interests is to ask questions.

Lots of questions.

To the point of being (mildly) annoying. Seriously.

How many questions do you ask? As many as it takes until you are confident that you can trust the company enough to work with them.

Your asking questions is actually one of the first tests to know whether it's a digital agency you can trust. An agency that is competent and aboveboard is going to welcome your questions and will have clear answers to them. An agency that is dishonest or doesn't have their act together won't have clear answers or will employ diversion tactics to avoid answering.

That being said, vetting an agency isn't just about asking questions in general but also asking the *right* questions. Think of this stage as a job interview for the agency. If you're about to hire this person or company to work on your behalf, asking the right questions will give you a clear indication whether or not the agency is worth hiring.

What are the "right" questions to ask? Ultimately, it differs from project to project. Different websites go through different stages of development, and not every good digital agency practices the exact same workflow. Even among the questions we recommend asking in this book, not all of them will be relevant to your business or your website project. The important guiding principle here is for you to *gain confidence* in a prospective digital agency, so the right question is any question that helps you gain that confidence. Likewise, there might be no single right *answer* to the question—you just need to be satisfied with the answer before moving forward.

Asking questions is so critical to hiring the right digital agency that, at the end of this book, you'll find a checklist of questions to ask before engaging an agency. These questions will help clarify your relationship with the agency once you've decided to work with them.

In this chapter, we're going to home in on a few key questions you should ask during the vetting process to help you determine whether to entrust an agency with your project. The answers you receive should tell you whether to move ahead with the company or walk away.

Have you done work for companies similar to mine? If so, may I see it?

The obvious idea behind this question is to determine if the digital agency has any experience working with a business like yours. If they have, take a look at their work, observe the design and flow of the site, and see if their solutions represent a good fit for your type of business.

It's not necessarily a deal breaker if they haven't worked with companies similar to yours, but it's something you'll have to take into account. If you're a law firm, for example, and the agency's portfolio consists mostly of drop-ship ecommerce sites and retail stores, you'll need to be prepared to navigate a pretty steep learning curve with the agency as they adapt to your business needs. If you're willing to walk that path, fine. If not, you might want to keep looking.

Is all the work being done in-house, or will you be subcontracting work to freelancers or overseas?

Some digital agencies handle all the aspects of development and design with full-time employees, while some may hire subcontractors to carry out certain elements of the project. For example, if you need web copy written for your site and the agency doesn't have a staff copywriter, they may hire a freelance writer to compose the copy for you.

It's not "right" or "wrong" if an agency hires subcontractors—many excellent websites are created with the work of highly experienced freelancers. This question is more about knowing who will have a hand in representing your business online. Ultimately, it's the agency's responsibility to make sure everything is correct with your site, whether they subcontract certain parts of it or not.

That said, it's completely appropriate for you to ask them to identify which elements of the site will be handled by freelance labor and who will perform the work. If you're uncomfortable with the proposed arrangement, you're free to look elsewhere. Likewise, if you feel their answer to this question is vague, it may be a warning sign.

What are the terms of payment?

Logistics and details about payments will be mapped out more concretely during the contract process, but it's important to bring this up during the vetting process, too. Why? Not only will it prepare you financially, but it will also help you to further vet the agency's professionalism.

Does the agency expect payment installments by certain calendar dates, for example, or by completion of certain project milestones? Or perhaps they tell you something vague, like, "Oh, don't worry about that right now—we'll get to that later."

If an agency can explain their payment expectations clearly, they've obviously put some effort into developing a payment policy that is fair to both parties, however they choose to be paid. Incidentally, you should *never* pay in full up front because it instantly makes you vulnerable by assuming all the risk, while the agency assumes none. If a digital agency asks for up-front payment in full without a convincing reason, this is a huge red flag.

Is this a fixed-price project? If not, what could make the price increase or decrease?

Bear in mind that you're not necessarily demanding that the digital agency give you a firm quote here—that can come later. What you're asking is whether the agency can give you a firm estimate once they understand the full scope of your project, and if not, what the variables might be. Your goal in asking this question is to determine whether the agency actually has developed consistent, fair pricing for its services, and whether you could be blindsided with unanticipated expenses further down the road.

What is the estimated time frame for a project like mine?

Once a good agency understands the scope of your project, they should be able to provide an estimate of how long the project will take, and you should be okay with committing to that time frame. Again, there's no litmus test for how long it should take—that depends on how much work must be done on your website, the agency's current schedule, etc.—but the key is that they should be able to provide *some* sort of specific answer, even if the timeline needs to be adjusted along the way. If they can't give you an estimate for completion, it's a signal that they might not have thought it through, or worse, that they don't have a good grasp of their own project management. If you work with the agency under those circumstances, be prepared for delays.

When my website goes down, who do I call or email? How long before I can expect a response?

We'll elaborate on this later, but statistically speaking, your website *will* go down at some point. By asking your digital agency what their procedure is in such cases, you're really asking whether they actually *have* a procedure in place. As long as they provide an answer you're okay with, you should be good to go. But if they balk at this question, expect long periods of downtime with slow response times because they're obviously not prepared for the inevitable.

NOT JUST *WHAT* THEY ANSWER, BUT *HOW* THEY ANSWER

The answers to all these questions aren't as important as *how* the digital agency answers them. *The most important thing is that the agency can answer the question, clearly and succinctly.* From there, you decide whether you can live with the specifics. If the answers are too vague and you can't get the agency to be more specific, that could be a red flag. Likewise, agency representatives acting annoyed or frustrated by your questions is another potential warning sign.

Your website *will* go down at some point.

Why is this so important? An experienced digital agency knows what can go wrong in a client relationship because chances are they've been through a few mishaps before you came along, and they have learned from their mistakes. A good agency *wants* to make the working relationship as clear as possible, so they will welcome as many questions as you need to ask.

ASK YOURSELF SOME QUESTIONS, TOO

So the digital agency has survived all your questions and answered at least most of them. Now it's time to ask yourself a few questions before moving forward.

Do you like their work?

When you look through the agency's portfolio, do you like what you see? It's not enough that you have a friend who strongly recommends them, or even that they conducted themselves in a professional manner. If you can't honestly say that you personally like at least 75 percent of their work, chances are you're not going to be happy with the work they do for you. This is highly subjective, of course, but it matters. They might be entirely trustworthy, but that doesn't mean they're the right fit for your business.

Do you respect the company?

In addition to liking their work, do you feel you can respect the people on the team, the agency's process, and the product? This is important to ask yourself, because you may be working with this digital agency for some time to come, especially if they will be hosting and maintaining your website after building it.

Can you accept the estimated time frame the agency gave you?

Most clients ask one question: "How long do you think this project will take?" It's a valid question, but not the right one. Most projects have multiple milestones. Heck, even washing your clothes has multiple milestones.

The question you need to ask is "How long will each phase take?" This will give you a much clearer idea of the microphases of your project—and if one phase runs long, you should expect the final deadline to do the same. Only airplanes are able to make up time in the air.

Are you satisfied with the digital agency's answers?

During the vetting process, were you in agreement with the agency's responses? If not, are there any deal breakers in the mix? Is there room to negotiate with the agency? If you have misgivings about the payment schedule, for example, can you suggest an alternative approach? In many cases, the agency might be flexible with certain policies in order to make the client comfortable. In others, they might stand firm. If you feel comfortable with most of the answers they give you, it might be worth it to ask about one or two sticking points. If you find yourself having trouble with most of their policies, however, it's probably best to walk away rather than press a list of issues.

CHECK THE NONREFERENCES

We have one final tip before we move on.

Perhaps you've noticed that one notably absent question from our list is "Can you provide any references?" I've omitted this question on purpose, not because you'd be remiss in asking for references from a digital agency, but because these references won't tell you the whole story. Almost every agency has at least a *few* satisfied customers, and it's a sure bet that every reference they give you will screen positive.

If you really want a better picture of how the agency works, look for the nonreferences—people who've worked with them who *aren't* on the agency's list of references. Do you have any mutual acquaintances who have worked with the agency? Can you find businesses online whose sites were built by them and call those businesses for more information?

Many agencies will place a backlink in the footer of a website they developed. These backlinks can help the agency improve their search engine rankings, but most importantly, they also serve as a visible symbol of pride in craftsmanship for the digital agency. A number of tools exist to help you find sites that link to your agency's website. Here are just a few:

- Monitor Backlinks, monitorbacklinks.com
- Open Site Explorer, moz.com/researchtools/ose/
- Majestic, majestic.com

You may also find reviews from businesses who have worked with the agency on sites like Google My Business, Yelp, and Clutch.co.

Take online reviews with a grain of salt, though. A one-star review isn't necessarily a deal breaker. It's the Internet and anyone can say anything. However, if there are consistent complaints among the agency's clients, you may want to listen to what they have to say.

CHAPTER 4

Choosing a Digital Agency: Red Flags to Avoid

"Ninety-three percent of communication is nonverbal."

Ever heard this statistic? It's a number that is often thrown around—and just as often debated. It comes from a famed 1972 study by psychologist Albert Mehrabian, who conducted multiple studies that concluded human communication is divided as follows:

- Words: 7 percent
- Tone of voice: 38 percent
- Body language: 55 percent

Add tone of voice and body language together, and we come up with 93 percent nonverbal communication.

Many scientists over the years have challenged these numbers, and even Mehrabian himself agrees that this formula shouldn't be taken as an absolute since he based his studies on certain controlled criteria. However, at the very least, most of these folks agree that nonverbal communication is paramount, and the vast majority of communication has little to do with actual words.

So what does this have to do with creating a website?

To put it simply, when you're vetting a digital agency and they're answering your questions, *don't just listen to the words.* Keep your eyes and ears open, because there is more being communicated nonverbally than there is verbally. An agency could provide all the right answers in theory, but you could still wind up regretting the decision to work with them because you weren't paying attention to nonverbal signals that might have tipped you off. Ignoring these signals could lead you to miss an important red flag.

Here in the Wild West of Internet-land, recognizing and avoiding red flags can be the key to protecting yourself against a bad deal, or at least a bad experience. Not every red flag is an automatic deal breaker, but it should at least prompt you to dig a bit deeper and ask some more questions. If the agency can't satisfy your concerns regarding the red flag, it's time to walk away.

In my experience, there are a few red flags that many people sail by when interviewing prospective agencies. Below you'll find a few bright-red flags you shouldn't ignore. If the phrase caveat emptor ("buyer beware") ever would apply, it'd be now.

VAGUE ANSWERS

When you ask a digital agency a question—especially one of the key questions we bring up here in the book—the agency should be able to give a clear, detailed answer. To illustrate, let's look at one of the simpler questions from Chapter 3: "If my website goes down, who do I call or email? How long before I can expect a response?"

If you ask this question, the digital agency should be able to respond with something specific like, "Bill is our tech-support guy; you can email him at bill@superduperwebagency.com and he'll respond within 24 hours." Or "You'll have 24/7/365 tech support at 1-800-XXX-XXXX." A specific answer like that is a good sign that the agency is on their game.

But let's suppose the agency responds with "Well, you shouldn't have any trouble, but if you do, call our office and we'll figure it out for you." That answer might sound reassuring, but it tells you absolutely nothing—it doesn't offer any solutions in the event your website goes down. If the digital agency can't give you a clear, concise answer to this simple question, make a note of it and keep listening. If you continue to receive vague answers like this in response to other questions you ask, you've got a clear red flag.

THE PEBBLE IN YOUR SHOE

When you're vetting a digital agency, you're effectively interviewing them for a job. Act as if you were hiring an employee to join your staff. The interview questions might be a bit different, but the process is about the same. In a job interview, you're not just checking the candidate's qualifications, you're also feeling out the employee to see whether they would be a good fit for your team. The same goes for a digital agency you're thinking about hiring.

If the behavior exhibited by the agency rubs you the wrong way during the initial interview, you're only seeing the tip of the iceberg.

I've interviewed and hired many people over the years. Some of those interviewees were bright, engaged, and dressed the part, and these folks have been great assets to my team. I've also had candidates who wouldn't look me in the eye, laughed way too loudly, or reeked of a cologne that could fend off attackers. I've learned by experience that if something about the candidate rubs me the wrong way during the interview, that trend is likely to continue and get worse if I hire that person. In other words, while I try to be as open as possible, I've also learned not to ignore the little indicators that could detract from our office environment.

The same applies to any digital agency you are considering. If the behavior exhibited by the agency rubs you the wrong way during the initial interview, you're only seeing the tip of the iceberg. It's like feeling a pebble in your shoe. It only bothers you a little at first, but if you don't take off your shoe and remove the pebble, it's going to feel excruciating after about a half mile of walking. If something annoys you or makes you uncomfortable at the beginning, you'll be even more uncomfortable the longer you let it go.

NONREFERENCES

At the end of Chapter 3, I mentioned it's a good idea to look past the references the digital agency gives you. Let's elaborate on this idea a bit.

Most digital agencies will have a curated list of references, and if you call them, they'll likely have great things to say—otherwise, the agency wouldn't have them on the list. But if you really want to dig deeper, find some companies they've worked with who *aren't* on that list. You can often find these through backlinks (mentioned earlier), mutual acquaintances, or even the sample websites the digital agency has provided for you. Call some of these companies and ask about their experience with the agency. If they left, find out why. If they are still with the agency but not on the list, you might find out what the unfiltered story is.

Now, while your hope is obviously to hear nothing negative about the agency, in the interest of fairness, a negative review shouldn't be a deal breaker in itself. There are many reasons why businesses make and break alliances, and there are also two sides to every story. If you do hear something negative, remember to take it with a grain of salt, and use your own best judgment. However, if you hear a significant number of horror stories from this uncurated list of nonreferences, or if several of them have the same complaint about the company, there's a pattern of broken relationships, and that's a red flag that you should not ignore.

YOUR GUT TELLS YOU NO

Admittedly, the gut test is highly subjective. But let's provide some context.

Suppose the digital agency you're considering is a referral from a trusted friend. This friend of yours works with the agency, loves them, and swears by them. However, after talking with the agency, you figure out that your friend's referral is the *only* positive thing they have going for them—nothing else jives. They give vague answers, their contractual agreement is one page long and says almost nothing, and the nonreferences you find aren't giving them good reports. In fact, if it weren't for your friend's strong referral, you wouldn't even be considering this company at this point. You feel like you're being biased and unfair. Is your friend seeing something in this company that you don't see?

No. Actually, the opposite is probably true: you're seeing something your friend *isn't* seeing. Your friend is saying yes, but your gut is telling you no.

Go with your gut.

In fact, this holds true even if you *don't* find overwhelming evidence that contradicts your friend's recommendation. You don't owe it to your friend or anyone else to enter a working relationship that you feel could harm your business. If you can't get comfortable with the agency, no matter how subjective that feeling is, there's no harm or foul in walking away. But let's also be clear: if all other signals except for your friend's recommendation are pointing in the opposite direction, you shouldn't walk away. You should *run*.

Trusting your gut is far from scientific, but not every business decision you make has to be. You always have the right to say no.

♦ ♦ ♦

Before moving into some of the more legal and technical aspects of working with a digital agency, let's address the most important question you can ask. By design, the answer to this question could be either the deciding factor for you, or the biggest red flag of all.

CHAPTER 5

The Most Important Question

In the last two chapters, we've discussed key questions to ask when you're thinking about hiring a digital agency to build your website, along with red flags to watch for and avoid. But there's one more question you should ask, one that stands above all the others, so much so that it warrants its own chapter. Your final decision to work with the digital agency should hinge on the answer to this question. If the agency does not answer it to your *complete* satisfaction, you should walk away and keep looking. Period.

Ready? Here's the question: "What is your cancellation policy?"

That's right. The most important question to ask any digital agency is effectively "What happens when we part ways? What happens when we break up?"

Because you *will* break up.

EVERY RELATIONSHIP ENDS

In polite society, we avoid talking about when things will end, especially when your relationship is just beginning. But in reality, every relationship ends.

Whether it's a business relationship lasting a few years, or the happiest, longest lasting marriage you can think of, everyone is going to part ways at some point, whether through dissolution or death. You're starting a new relationship with this digital agency; believe it or not, the healthiest way to do that is to talk about what happens when the relationship ends. Think of it as sort of a prenuptial agreement: *when* you break up (not *if*), who gets what?

Granted, if you tried to start a *romantic* relationship this way—well, let's just say you might want to plan on staying single for a while. But in a business relationship, it's actually one of the best ways to preserve goodwill on both sides. If your digital agency is uncomfortable even broaching the subject, that's a red flag.

THE BENEFIT OF PRIOR EXPERIENCE

A professional will have had clients close their account, and thus they will have a proven process by which the client leaves their agency. A good digital agency won't evade this question, because they realize many of the people seeking them out have already had to deal with breaking up a rocky relationship with another agency. If their answer is confident, cool, and assured, you know you're dealing with a professional. They've ended relationships with other clients, and they've acted professionally.

However, if they jokingly laugh off the question with a "Why would you ever want to leave us?" spiel, then be prepared for a messy breakup down the road. If they make inflated promises about how they'll ensure you never want to leave and bend over backward to keep you, chances are you're not dealing with a reputable agency. They might have good intentions, and there may be no scam involved, but when you choose to leave that company (again, not *if*, but *when*), it's going to be a sticky situation because they don't have a process in place. You'll have to ride the learning curve with them should you ever decide to leave.

BUT WHAT IF THE DEAL IS REALLY GOOD OTHERWISE?

Some deals may seem too good to pass up, so much so that you might be tempted to risk the messy breakup. If you're willing to assume that risk, more power to you. I'm not saying you should automatically run away if the digital agency doesn't have a set cancellation policy. All I'm saying is you need to be *satisfied* with the answer. Mark my words, you'll get to put their policy to practice one day. If the cancellation policy isn't clear but you trust them otherwise, you simply have to weigh the benefits against the risk, and be satisfied with your choice.

DETAILS OF THE CANCELLATION POLICY

Optimally, when you ask the cancellation question, the digital agency should provide an initial response that puts you at ease—something like, "If you cancel with us, we'll hand over our work product, and we'll cooperate with your new digital agency to help make the transition seamless." This is the best sort of answer you can hope for, because once again, it's a signal that the agency has been around the block.

Here are some questions to ask before entering the engagement, so you'll know how the relationship will end:
- What happens to the money already invested? Is there a full or partial or any refund?
- How is a partial refund calculated? By time spent, or by milestones reached?
- Can either side trigger a cancellation? If so, how?
- Are there "exit points"—that is, specific points in the development process where cancellation can occur? And what, if any, penalty is there?
- When we part ways, what becomes of the work, such as the design files and the website? Do the files belong to the agency, or can you take them with you?
- What happens to the website if the digital agency goes out of business? How do you access your files?
- What happens to the website if the digital agency gets bought or merges with another agency?

◆ ◆ ◆

In Part Two, we'll venture into more detail about the specifics of a good cancellation policy and how it should be spelled out in your contract. For now, just be aware that without this provision in your contract with a digital agency, you are putting yourself at risk.

PART TWO: THE **LEGAL STUFF**

CHAPTER 6

The Fine Print: Know What's in Your Contract

Let's be honest about contracts: no one wants to read them. No one gets excited about the legalese contained in them. Blah, blah, blah. It's all standard procedure, right? Just sign on the dotted line, and let's get on with it.

No one cares about the fine print—that is, until after they get burned by a bad company. Then they're very interested. Of course, by that point, it may be too late to do anything about it.

Ultimately, if you have a dispute with a company or wind up getting burned while trying to build your website, chances are it's going to be over something that was in your contract or wasn't but should have been, if there was even a contract at all. That's why whenever you deal with a digital agency—or for that matter with any company—take the time to read the contract *before* you sign it, not after. Yes, contracts are boring, but they are essential to any healthy business relationship.

All too often prospective clients want to jump to "Let's look at websites we like" rather than first discuss the terms and scope of work needed. It's easy to say what you like; we all have opinions. And shopping for a new website design can be fun. However, if the contract isn't clear to both parties, the project won't be very fun at all.

Why is the contract so important? First and foremost, it defines the relationship between you (the client) and the company: what the agency is going to do for you, by what time, for how much, etc. It also details how you'll move forward if something goes wrong. Second, the contract details your rights and responsibilities with regard to the digital agency. In other words, it lays out what you need to do in order to help the agency fulfill their part of the bargain, as well as what you are entitled to from the agency. Finally, the contract protects you legally if the agency fails to meet the terms of the agreement.

At least, that's what it's supposed to do. If the contract is vague or fails to address certain key elements of your relationship with the agency, it can leave you unprotected and vulnerable. In fact, you're more likely to get screwed over by what is *not* in the contract than by what is. This is why, as boring as it might seem, you must not take your contract lightly. Most importantly, you should not sign the contract until you are confident that you fully understand it and what is included—and, often just as important, not included.

Take the time to read the contract *before* you sign it, not after.

Every contract is different, of course. A contract for a simple website project doesn't necessarily have to be as extensive or detailed as a contract for a more complicated one. However, there are a few basic elements that should be addressed in any contract you sign with a digital agency. Let's go over these now.

OWNERSHIP OF THE WEBSITE AND ITS CONTENTS

One of the most important, and often overlooked, things to address is ownership. Once the developer has completed the project for you, who owns the work? Ordinarily, you might assume you own the website and all its contents, and in many cases that's true. However, there are some important caveats to look for.

Work for hire

A *work for hire* is when the entirety of the product developed belongs to you. It can include the files used to design the site; associated graphics, videos, and stock images; and the database programming. This is the simplest form of ownership, but it can also be the most expensive, because all the elements of the website must be built from scratch in order to be transferred to you in the event that you are no longer a customer.

License to reuse software

In many cases, there may be limitations in your agreement as to the level of ownership offered. The digital agency might have developed their own hosting platform or other software elements, and in the contract they state that they're granting you a license to use the work product as long as you make recurring payments. This isn't a deal breaker, but you need to understand that this provision ties your website to the agency for ongoing services. You may own the content on the site, but if you leave the digital agency and stop making the payments, your services may be suspended or terminated. So just be clear about what type of product you want and what type of product you're purchasing.

Software as a Service (SaaS)

In some situations your digital agency might utilize a set of licensed products to build and manage your website. These products are referred to as Software as a Service, or SaaS, and may include products like Wix or Squarespace. For all intents and purposes, these are leased websites that are owned by the software company. When you contract with one of these services, you're paying a fee for a license to use their product that is revocable. If your website violates their Terms of Service (ToS), if they dissolve their business, or if you discontinue paying their recurring charge, the website you worked so hard to build on their platform is gone.

When a digital agency you've hired utilizes a SaaS, the following are a few very important questions to address in your contract with the agency:

- Who owns the software license associated with your website? In other words, is it your organization's license, or the agency's?
- If there are fees for the use of the software, who makes the payments month after month, year after year? Does the agency make these payments for you and pass the cost to you? Or are you expected to pay the fees directly to the SaaS provider?
- If the agency pays the fees but goes out of business, what happens to your website when they stop making the payments?

Other licenses

Beyond the content management system, or CMS, that your website runs on, there are other elements of your site that might require licensing. For example, many stock photos are licensed for use for one year at a time. If your site uses a shopping cart, there may be a separate license for the shopping cart software, the security certificate, the merchant account services, and the web hosting.

If additional licenses are required to build your site, who holds those license keys, how much is the renewal, and who will be responsible for the renewal of those licenses? Is it you, or the digital agency? All these details should be spelled out in the contract to avoid any pitfalls down the road.

In most cases, the written content becomes your property, although in rare cases some of the content could be licensed from your digital agency or a SaaS. This question is particularly important for stock photos and graphics you might use on the website. Who owns the license to use them? Do you have a license of your own, or are the photos posted under the developer's license? Even worse—is there even a license at all?

This issue gets people into trouble more often than you think. I always recommend that my clients open their own account with the stock photo provider so they can purchase the license themselves and, if necessary, manage any renewals. If they forget, stock providers are very good at sending out cease-and-desist orders and fine letters for violations. Let me give you an example of what can happen when the license to use stock photography isn't properly cleared.

A digital agency we know had a client in need of a website. At the time, they were subcontracting the graphics work to a graphic designer who was doing a limited stint with the company. Unbeknownst to them, this designer put together a web design that included images they had found on Google Images and just saved without purchasing the rights. Those images were not the client's property, nor did anyone have the authorization to include those images in the design. The client loved the design, and the digital agency was none the wiser.

Six years later, the digital agency received a cease-and-desist request from an attorney who represented the owner of one of the images that had been used without authorization. Accompanying the request was an invoice for $850 for unauthorized use of the photo, offered as an alternative to litigation. Sadly, had the graphic designer done their job properly, the agency could have licensed the stock photo six years prior for maybe five bucks. But because of the historical nature of the web, the copyright owner could easily prove that their image had been used illegally year after year. The graphics subcontractor was long gone by this time, and the unauthorized use wasn't the client's fault—so the agency was left to cough up $850.

In this case, the digital agency, not the client, got burned by a dishonest or perhaps just careless subcontractor. But situations like these underscore the importance of respecting and protecting *all* intellectual property. To keep things simple, it's best for you to purchase your own license to any stock photography so it's not tied to your digital agency, but in any case, always be certain that the contract spells out who is responsible for ownership and licensing of the photos and graphics. If your digital agency owns the license, that's fine, but you'll need to replace those photos or buy your own license if you ever part ways with the agency.

STATEMENT OF WORK AND SERVICE LEVEL AGREEMENT

Now let's talk a little bit about Statements of Work and Service Level Agreements. Depending on your needs, both may be important aspects of your contract with a digital agency. The Statement of Work (SOW) details all your project deliverables and their deadlines, while the Service Level Agreement (SLA) covers the terms of any ongoing services provided by your digital agency, such as web hosting, site maintenance, etc. If you're hiring the digital agency only to build your website, you'll need just the SOW, not the SLA, since the agency will not be providing services after the engagement.

We'll cover these terms in more detail in the next chapter; for now, just know that, depending on what you're asking the agency to do, one or both should be included in your contract before you sign. Otherwise, if the agency fails to meet your expectations, you won't have a way to prove the terms were not satisfied.

SATISFACTION

While we're on the subject, we need to talk about the word *satisfaction* and what it means in your contract. At what point are the terms of the contract satisfied?

In your contract, payments may hinge on the completion of certain milestones. But to whose satisfaction? Is it solely within the digital agency's authority to determine when certain requirements or milestones in the project are satisfied? Or is that the customer's call? And what, exactly, constitutes satisfaction?

For example, the SOW may say that you will get two rounds of revisions. In other cases, the project requirements may be satisfied after the agency has provided a set number of hours of labor. And I've seen SOWs where the project's terms would be satisfied only after the client's senior management signed off on the final work product.

In the contract, *satisfaction* might refer to the client's satisfaction, or it might mean the agency has fulfilled certain criteria. Just make sure the terms of satisfaction are clear in the contract; otherwise, it opens the door for disputes.

WARRANTIES AND LOSS

Another key element that should be in your contract is a section about warranties. What type of warranty does the digital agency offer regarding your website and/or their services, and how long does it last? Furthermore, how does the contract and/or warranty protect you against losses that might occur if the agency fails to do their job? We'll cover this ground more fully in Chapter 9, but just be aware of it for now, and know that it's important that your contract not omit this section.

THE BREAKUP

As we discussed in the previous chapter, no matter how solid your relationship with the digital agency is, or how much you like their work, at some point you're going to part ways. Either you'll need a new website and want to try a new agency, or you or your agency will merge, sell the business, or close altogether. Or you might find during the development process that your digital agency (or their software) is intolerable.

One way or another, it's not a matter of *if* you're going to break up, it's a matter of *when*. So you need to plan for it, and the terms of the breakup need to be spelled out in the contract. We won't rehash here all the details covered in Chapter 5—but bear in mind that every good contractual agreement will detail how, when, and for how much the services can be cancelled, so when you part ways, you know what you'll be parting with.

Kill fee

Every contract should contain a provision for what is called a *kill fee*. If, after accepting the agreement, you or the digital agency wish to part ways, the kill fee section in the contract states how the agreement can be terminated and details what happens to the fees that have already been paid. For example, will the digital agency refund some or all of that money?

What constitutes a good kill fee section depends on the client and the agency and their respective comfort levels with financial exposure. In other words, how much up-front investment are you willing to part with, and how much work is the digital agency willing to absorb without getting paid? By making sure this provision is spelled out in the contract, you can help avoid the messiest of divorces.

Pause clause

In some agreements, there's a *pause clause* in addition to the kill fee. The pause clause allows you to temporarily stop the process of producing your website for a specified length of time, either without penalty or perhaps for a modest fee. This provision is helpful because it suspends the relationship rather than terminating it completely. So if, for example, unexpected circumstances prohibit you from fulfilling your part of the project by a deadline, or if your company experiences a temporary budget shortfall, you can effectively hit "Pause" on the website project until the circumstances right themselves. This is not a must-have provision by any means, but it can be helpful if you need a bit of flexibility and if the digital agency is willing to work with you.

◆ ◆ ◆

As you can hopefully see by now, hiring a digital agency to build you a website is a bit more involved than "Here's some money, please make me something nice." There are many details to consider in establishing a web presence, and as a result, there are many areas where things can go wrong. This is what makes the contract such an important part of your relationship with any digital agency. An agency that is trustworthy will take this part of the process very seriously, while an incompetent or dishonest one will gloss over it.

The bottom line is that before you hand the reins over to any digital agency, make sure you have a contract, make sure you understand what's in it, and make sure you agree with it. To do anything less is to court trouble.

CHAPTER 7

Understanding Statements of Work and Service Level Agreements

In the last chapter, we discussed several elements that should be included in your contract with a digital agency. I want to elaborate more on two of them: the Statement of Work (SOW) and the Service Level Agreement (SLA). These documents could represent the entirety of your contract, or they may be contained within it. Likewise, some projects only require an SOW or an SLA, while others require both.

Either way, these two documents describe specific aspects of how the agency is going to serve your needs as a client, so you need to understand what they are and what they should contain. Since the SOW and SLA represent two different facets of your working relationship with the digital agency, let's go over each of them separately.

STATEMENT OF WORK

The Statement of Work (SOW), or Scope of Work, provides a detailed and descriptive list of all the deliverables of your website project—in other words, the specific features and functionality the digital agency is going to develop for your web presence—and when they should be completed. The SOW is important, not only because it tells you what to expect from your digital agency, but also because it serves as a compass for project management so both parties can track the progress of the website project and make sure everything is going according to plan.

A well-constructed SOW will be very detailed—sometimes painstakingly so. Why? Because a vague SOW only opens the door to disputes. It's in the best interest of both parties to eliminate vagueness whenever possible.

> **A vague SOW only opens the door to disputes.**

For example, a statement to the effect of "The agency will construct a website for the client by November 1" is far too ambiguous. It fails to address how many pages the website will have, what features will be included, what constitutes a completed website, or how the client's satisfaction fits into the equation. Under this vague language, the agency could technically develop and upload a one-page website on October 31 with a few words describing your company, claim the terms of agreement have been satisfied, and walk away with your money. There would be precious little you could do about it.

So even though it may feel tedious, make sure the language in the SOW is as descriptive as possible. Leave absolutely no room for misinterpretation. To facilitate this goal, we'll now cover the elements that a good Statement of Work should contain. It may cover other elements than those discussed—for example, a section for security or other special considerations—but the points discussed below provide a good basic framework for an SOW you can trust.

Scope of work

This section contains the list of concrete deliverables—the specific items you're hiring the digital agency to provide for you. These deliverables may include any or all of the following:
- The website itself (including the number of pages)
- Specific features of the website that require additional effort (e.g., custom forms, responsive layout, ecommerce elements)
- Web hosting (a place to host your website online)
- Written content for the website
- A monthly maintenance commitment (whether the agency provides a certain amount of maintenance, and how many hours per month it includes)
- Post-launch marketing services

Work requirements

This section breaks down the entire project into specific tasks that will be required to complete the deliverables as outlined in the scope of work. For example, in developing a website, what tasks will be performed during the design phase and by whom? What happens during the build phase? The digital agency should detail these items here.

Time frame, schedule, and milestones

This information might be found in multiple sections (e.g., Period of Performance, Schedule of Completion), but somewhere, the digital agency should specify how long your project is expected to take, along with a detailed schedule and milestones for completing each step of the project. This information not only provides a roadmap and timeline, but it also helps you understand the development of your website step-by-step.

Acceptance criteria and satisfaction

This section discusses how you will approve certain steps in the design, build, and implementation of your website before the next steps are taken, and who is responsible for signing off on each step. This is also where you should find clarity on what *satisfaction* means, as we discussed in Chapter 6.

SERVICE LEVEL AGREEMENT

As we noted earlier, while the SOW mainly deals with specific deliverables associated with the development of a website, the Service Level Agreement (SLA) details the terms of any ongoing services provided by your digital agency, such as web hosting, site maintenance, etc. If you're hiring the digital agency only to build your website, and you plan to host it elsewhere, you'll need only the SOW. However, if you're entering an ongoing relationship with this provider that involves servicing, hosting, and/or marketing your site, make sure your contract paperwork includes an SLA, and make sure you understand its terms.

The SLA is a critical part of your contractual agreement, especially if the digital agency is hosting your site, for one very important reason: at some point, *your website is going to go down*. It is a computer-operated system, and all computers eventually fail—so when the computer that hosts your site fails, your website goes off-line until the digital agency fixes the problem or replaces the server. You need to know how the agency is going to respond when that happens and what happens to your website—and your business. (Again, it's "when," not "if.")

All computers eventually fail.

The SLA should contain certain basic components to protect your interests. Let's go over these now.

Scope of service

This section tells you exactly what ongoing service(s) you're hiring the digital agency to do. It will usually cover website hosting and maintenance, but it may also include additional services like updating your blog, email marketing, tracking metrics, etc.

Performance level and reliability

This part of the SLA details the level of service you'll be receiving—for example, how much bandwidth your website is permitted each month, how much uptime is guaranteed, how much downtime is allowed within the agreement, as well as what the terms are for off-line maintenance.

It is normal for a certain amount of acceptable downtime to be included in your SLA. Your web host may guarantee 99.9 percent uptime in their SLA. But taking into account the number of hours in a year, that means the web host is allowed to let your website go down for *four hours per year*. You may think a

99.9 percent uptime agreement is near perfect. However, you may not think it's perfect enough during those four hours of allowed downtime.

In addition, there may be exceptions to the downtime allowance provided by the SLA. For example, so long as the agency provides advance notice, overnight downtime resulting from scheduled maintenance typically isn't included in your four hours per year. That's right. You could face many hours of scheduled downtime even with 99.9 percent uptime.

Monitoring

How will the digital agency monitor its own service to make sure it's meeting the minimum requirements? Is there a third party that audits the uptime? This section has the answers.

Recompense for failed service

This section is very important to you as a customer, because it tells you in what ways (if any) you'll be compensated for any failure of service. For example, if your website goes down for longer than the acceptable amount of downtime, will you receive credit for that downtime? If you lose income as a result of that downtime, to what extent, if any, is the service provider responsible?

Reporting process and response times

This is the part of the SLA that outlines how long the digital agency has to respond to server failures or other problems afflicting your service.

• • •

The important thing to remember is that *anything not specifically mentioned in your agreement is not included in your agreement.* If you need something from your digital agency that is not put down in writing, the agency is *not* legally obligated to provide it. So if you need a particular service, a specified response time, 24/7/365 accessibility, or anything else to keep your website and business functional, make sure those terms are included in the SLA before you sign.

CHAPTER 8

Avoiding Scope Creep

Once the Statement of Work and/or the Service Level Agreement have been finalized, the goal of both parties should be to adhere to those agreements as much as possible. But what happens if you sign the agreement, then decide you need to add something to the proposed website that wasn't originally discussed? What if you decide you want to add a new ecommerce element or a blog?

Once the work has begun on developing your website, it becomes a project in progress. Making changes midstream is known as *scope creep*. If you initiate scope creep, you're now asking for things that weren't originally agreed upon, which adds up to more work and time spent for the digital agency. If the agency initiates scope creep, they are adding steps to the process that they didn't tell you about before. In the former case, it's essentially unfair to the agency; in the latter, it's careless on their part. Either way, it can add up to additional time and cost, causing friction in the process.

CHANGE ORDERS

One thing that can potentially damage the client-agency relationship is a change order. If a change order becomes necessary, the initial scope of work no longer fits the new requirements. This gets especially dicey when the client requests additional work and doesn't *recognize* it as a change to the initial scope, which can in turn lead to disputes over missed deadlines and rising costs.

I've seen change orders destroy the working relationship between the client and the digital agency fairly quickly, especially if the project had a vague scope of work that failed to address questions such as the following:

- How many pages will the website have?
- How many designs will it have?
- How many revisions are included?
- How many weeks will the project take?
- What happens if you are unhappy with the deliverables?
- What happens if you want to add something to the project? Is it included in this current development phase, or does it elicit a change order for addition as a second phase?
- How many forms does the website have?

There are a lot of different components that need to be thought out, and making decisions midproject is not the best way to do things. As the buyer of the website, you never want to be caught in a situation where you're saying, "Oh, I thought that was included."

COMMON METHODOLOGIES USED IN PROJECT MANAGEMENT

To better manage the issue of change orders, a growing number of digital agencies are implementing a methodology called *agile project management*. It essentially says, "We're going to allocate our team for a specified amount of time to work on your project, and these are the things we're going to accomplish during that time. And if you want to change your mind during that process, that's fine; we'll just need to add an additional period of time." This period of time in the agile development process is called a *sprint*. A sprint can last one or two weeks.

Agile project management addresses the issue of possible midstream changes upfront and figures them into the initial price of the project. In contrast, waterfall project management relies on a series of cascading steps, where the completion of one step leads to the next and the next. Under this method, going back to make unplanned changes can be hazardous to the project by triggering change orders, longer deadline extensions, and added costs. Let's say your site's designs have already been reviewed and signed off on, but then you come back and say, "The big boss looked at the website and didn't really like it—can we make some changes to the design?" This throws a hefty wrench into the process, because it means several steps will have to be redone in order to facilitate that change. And, unfortunately, those added steps often create added costs.

Think carefully about the style of project management that will work best for your project and how your organization typically makes decisions. If your budget allows for agile project management, it may give you some flexibility and help you avoid incurring dreaded change orders.

WHERE CHANGE ORDERS ARE LIKELY TO HAPPEN

From the aesthetic to the functional, change orders happen for all kinds of reasons. For example, maybe a client initially wanted to simply feature an online product catalog, but then decided after the initial scope that they would like to give customers the option to make purchases directly on the site. That's just one example, but really, change orders can involve just about anything—big or small:

"Whoops, we forgot to include our social media icons!"

"Now that I think about it, it would be really good to have a quote request form."

"Actually, this color scheme is not working for us."

"Oh, is it too late to include an email newsletter sign-up?"

"We're rolling out a brand-new product and need the website to reflect that."

Even with the most detailed contract, change orders like this can still pop up and create scope creep. But how did these needs manage to slip through the cracks in the first place?

WHY CHANGE ORDERS HAPPEN, AND WHY IT CAUSES FRICTION

One of the most common causes of scope creep is staff change. Perhaps you assigned your project manager Cindy to work with the digital agency. Then Cindy left your company or got pulled onto another project, and now Dave is going to replace her. The problem here is that Cindy already made a lot of decisions, and Dave has a whole different vision regarding how this project is going to be completed. Unless you and Dave have an agreement that Cindy's decisions must stand, Dave is likely to request some type of change that will end up increasing the scope of work, and by extension, the project cost.

Change orders and scope creep also typically occur when a decision-maker is AWOL during a critical decision, then swoops in later and asks to make a change. If a team or committee is in charge of making the decision, and not everyone signs off on it, this is basically scope creep waiting to happen.

> **If only a verbal agreement—and no documentation—exists to back up an expectation, it doesn't take long for things to get ugly.**

Granted, in our imperfect world, people are going to change their minds once in a while. Most agencies aren't going to give you grief because you want a change, especially if you're willing to absorb the extra cost and effort. Most of the time, the friction occurs when the client doesn't recognize a request as a change, and therefore doesn't want to pay the extra cost. If something isn't clearly spelled out in the scope of work, a he said, she said situation can quickly develop. If only a verbal agreement—and no documentation—exists to back up an expectation, it doesn't take long for things to get ugly.

So why are disagreements so common, if all you need to do to avoid them is spell everything out in the SOW or SLA? Most often the breakdown occurs when the salesperson hands off the project to the production team.

Let's say that you're a client in a sales meeting with Bob, the salesperson at XYZ Web Company. Bob writes up the quote, and everyone signs it and shakes hands. However, not all the information discussed in this meeting gets passed on to the production team. The handoff is an important moment. If any of the things you and Bob agreed to verbally are overlooked or forgotten, they won't be written down in the SOW, and, as a result, you risk a breakdown occurring later.

This is why it's so critical that you review the written agreement before you sign it, to make sure that everything you agreed to verbally with the sales consultant is included in your SOW. Mistakes may be unintentional—we're only human, after all—but this is where you as the client must be vigilant.

SLOW YOUR ROLL

By the same token, there is a chance it might actually be you, not Bob the sales guy, who forgot to mention something you wanted to include. After all, there are many details involved in the development and design of a website.

As a client, you can reduce the risk of change orders and scope creep by just *slowing down* when you're planning, discussing, and signing all the paperwork associated with your project. People often get so excited to start on their project that they overlook key details that need to be covered. Just take some extra time to come up with a list of every element you want to include in your website, and double-check to make sure everything you addressed is spelled out in the SOW before you sign. *Do not assume anything during this process.* Assumptions are another area where breakdowns can occur.

SCOPE CREEP ISN'T COMPLETELY PREVENTABLE

It's virtually impossible to entirely avoid scope creep. During a three- or six-month website-development project, minds are going to change. It's very rare to see a project launch with the same direction and goals it had on Day One. As time goes on, opinions and preferences inevitably evolve, especially when you're working with a team, or when new team members join the project.

Even with the most meticulous planning and detailed scope of work, there are some things you can't anticipate at the beginning of a project. To complicate matters, the larger the project is, the greater the likelihood that you will *want* to make some changes.

It's virtually impossible to entirely avoid scope creep.

Think about the last time you went into a big-box store like Costco and said to yourself, "I'm going to buy milk and eggs." You've budgeted three dollars for each item, so your *scope* going into that store is six dollars for two items. Now, perhaps you have the self-possession to walk into that store filled with unexpected deals and buy just those two things. But if you're anything like me, chances are you're going to walk out with more than just milk and eggs. Something is going to catch your eye or be on sale, or maybe you're going to see something else you remember you needed. And the longer you spend in the store, the more likely it is.

The same concept applies to developing a website, except in this case, you're basically spending three to six months in the store shopping. If you're having regular meetings with your own team members and your contact at the agency, you'll likely hear a lot of shiny new ideas being tossed around—ideas you hadn't thought of when you first decided to have a website designed.

The point is that even with the best planning and discipline, you can't anticipate everything. You want to keep your costs from creeping up unnecessarily, but you also want to allow for *some* flexibility, so if you want to make a change, it doesn't cause a crisis. To avoid getting caught off guard when this happens, allow space in your budget for contingencies.

Red flag alert: Catching scope creep

A good agency will actually call you out on scope creep when it happens. A bad one won't. If you're sending out-of-scope changes to your digital agency, and they don't point out that those changes aren't included in the terms of the agreement, that's actually a sign of an unhealthy company. When the agency isn't strong enough to tell you no, that means they don't have healthy boundaries; and if they're doing this with you, they're probably doing that with other clients as well.

A healthy company will address changes like these upfront by saying, "This was not in the terms of our agreement; we're happy to do it, but it will cost more money." This demonstrates that the company is willing to protect both your agreement and their product. Because *changes usually cost the agency money*. So, if they aren't passing those costs to you, they're absorbing them, and they can't do that for very long and remain solvent. The agency might be good enough to give you a freebie now and then, but if you're making a lot of changes and they aren't charging you for them, don't expect that agency to be around for long. And if the agency folds, where does that leave your website?

A good agency will actually call you out on scope creep when it happens. A bad one won't.

Another thing to watch out for is unexpected charges for change orders. If the agency does charge you for the changes but doesn't communicate this to you upfront, you could be hit with a "gotcha" bill at the end of the project. To protect yourself, make sure there is a provision in your agreement that says, "The customer shall not pay any invoices that are not pre-approved." That way, by virtue of the agreement, the agency has to inform you of additional charges before the work is performed, or it is assumed that the work is being done free of charge.

CHAPTER 9

When Things Go Wrong

I hate to be the bearer of bad news, but here goes: despite the best-laid plans, something will eventually go wrong with your website.

It's true, even with the best of digital agencies. It's an imperfect world, with imperfect people and an imperfect Internet. No matter how carefully you screen the agency, no matter how skilled they are, no matter how well you work together, no matter how ironclad your contract is—it's only a matter of time before something goes awry. The servers are going to crash, and your website will go down, possibly during a key sales time, costing you potentially thousands of dollars in sales. Web standards will change, rendering certain key features of your website nonfunctional. The digital agency will change hands, and someone will lose track of your assets in the process.

You get the idea.

I'm not trying to discourage you. I point this out because failures are a reality of the Internet (and life), and when you experience one, it is *not* necessarily a sign that you are working with a bad digital agency. Even with the best designed websites and the best agencies managing them, the probability of something eventually going wrong is near 100 percent. The key to weathering these storms is not to expect or demand perfection, but rather to have provisions in place within your contract to deal with these issues when they occur. A good contractual agreement will try to anticipate the most common types of issues and prescribe a remedy for each of them.

To that end, let's take a look at some of the elements in your contractual agreement that specifically deal with different types of website failures, and what is fair recourse when these things happen.

Even with the best designed websites and the best agencies managing them, the probability of something eventually going wrong is near 100 percent.

WARRANTIES

In Chapter 6, we told you that you'll want to be sure your contract with any digital agency includes a section on warranties. Now let's look at why—and what your warranty should, ideally, cover.

Let's say you've launched your website, and everything looks great. But two weeks later, you discover a bug, something that isn't working correctly on the site. This could be something small—for example, a button that doesn't work—or it could be something bigger that prevents the website from functioning properly. I specifically say two weeks in this scenario, because if something is wrong with the website, you'll know about it in the first few weeks. Either part of the site will stop working for some reason, or a fresh set of eyes is going to find something. The big question is who is on the hook financially to get it fixed? This is where your warranty comes in.

It's unreasonable to expect a lifetime warranty.

Warranties for websites may vary greatly in terms, so it's important to read over yours to know (a) how long it lasts, and (b) what it covers. There is no set standard, but knowing in advance what your warranty covers and for how long can avert lots of finger-pointing later on. It's unreasonable to expect a lifetime warranty, but make sure you're comfortable with the warranty terms before signing your contract with the digital agency, and ask questions if something is unclear.

Future functionality

One of the advantages of building your site with a Software as a Service website builder is that it will very likely have routine updates that keep your site running smoothly. However, if you decide to hire a digital agency to assist you, you may not be covered when technology advances.

With a constant and growing threat of hackers, web servers and content management systems are constantly being patched and upgraded to improve security and performance. However, some of these patches and upgrades can prevent code that previously worked well from operating entirely. Code can commonly become deprecated (out-of-date) and no longer work as it once did, all because the hosting environment was made more secure. You'll want to ask if your website includes a warranty for upgrades that will inevitably be needed at some point.

Let the warranty fit the project

It's appropriate to ask for longer, more detailed warranties for longer, more detailed projects. For example, if you have a particularly large website with lots of features, a sixty-day warranty might be insufficient because it might take longer than that to find all the bugs. Likewise, more complex websites are more likely to suffer compatibility issues as servers and protocols are upgraded, so it would be helpful for the warranty to cover these issues for a time. Regardless of the terms of the warranty, make sure you know and understand them and can agree to them.

COVERING LOSS

Because it's inevitable that *something* will go wrong, it's important to know what your contract with the digital agency says if that something results in a loss to your business. Let's say, for example, you run a website that sells cupcakes online. Each day, you sell an average of three thousand dollars worth of cupcakes. Then your website goes down, maybe due to a web hosting outage, a bug, or any number of things. For each day that the site is down, you're now sustaining a loss of three thousand dollars. Who is going to absorb that? Is the developer held harmless? Does the agency's insurance cover your loss of income, and under what circumstances? Can you, as a practical matter, bring a claim against the digital agency if the downtime is their fault? Or does your contract limit you to arbitration or filing suit in another state? Do you basically have to eat those cupcakes (pun intended)?

In most web-hosting contracts, a certain amount of downtime is permitted without penalty. As was mentioned before, if your web host guarantees you 99.9 percent uptime, this translates to *four hours per year* that the website is allowed to go down without recourse—and this usually doesn't include planned downtime for maintenance. Translated into your cupcake sales, if downtime costs you three thousand dollars per day, you should expect a minimum planned loss in sales of five hundred dollars per year under those terms. Beyond this, if your sales are dependent on website uptime, you should look for any other type of provision in the contract describing who's liable and what happens when certain issues occur.

WHEN YOUR AGENCY IS NO LONGER YOUR AGENCY

The business world is always changing, with companies starting, folding, and being bought and sold all the time. Sometimes changes in business relationships can have a direct effect on your website. A good contract will have some provisions in place so your web presence can adapt to these fluid changes, whether it happens on your end or on the side of the digital agency.

Let's suppose your digital agency is bought by another company, or worse, goes out of business. What happens to your assets? Does the website just go down, or do you have the option to retain your website assets and take them to another company? A good digital agency will work to keep your website intact through any sort of transition and will hopefully give you advance notice of any key changes. Of course, things don't always happen this way. What then?

There are so many scenarios that can occur here that it's impossible to anticipate a plan of action for all of them, so the key question to ask is "What happens to my website if the agency changes hands or closes?" If possible, there should be a safeguard in your agreement that allows you to access your files, photos, and content and take them elsewhere so that your website experiences as little downtime as possible. If your site is hosted on proprietary software, you might have to rebuild the website with another agency, but at least if the contract provides you access to the basic elements of your site in the event of closure, you'll have the building blocks to work with.

In many cases, a change of agency ownership won't affect your site at all, but it might affect the terms of your agreement. To account for this, look for a provision in the contract that requires the digital agency to notify you of any changes in terms and gives you the option of rejecting them.

WHEN YOUR BUSINESS CHANGES

Let's use the cupcake shop analogy one more time. Let's say you run your cupcake shop with a partner, and your partnership dissolves. Who gets the website in that case? This needs to be spelled out upfront, both in the contract you sign with your digital agency and in your business's partnership agreement. If your partner is branching out to start her own competing company, you don't want to have two identical copies of the website online. Only one party should retain ownership of the original site, or the site will have to be scrapped and built from scratch.

Again, there's no one-size-fits-all provision here, but the central questions to ask are "Who owns the original website?" and "Who takes ownership if the business splits?" If you purchase the site under a partnership or corporation, your agreement with the digital agency should match your partnership or corporation agreements, which detail what happens to the company assets, like a website, when your business relationship dissolves.

WORST-CASE SCENARIO: LITIGATION

If you're considering litigation, it's a sign that something has gone horribly wrong between you and your agency. If you find yourself in this unpleasant situation, your contract will likely include some language about settling disputes that will describe how to move forward. In particular, you want to zero in on two words that may appear: *venue* and *arbitration*.

Venue

Most legal contracts will specify a venue where lawsuits must be filed, and in most cases, it will be a courthouse in the same location where the digital agency is based. If you're working with a local company, this will likely be a nonissue; however, if you're working with a larger corporation, the venue might be more problematic. You might be required to travel to another city, state, or even another country to file suit. I've read some contracts for larger web companies that specify the venue as New York City, London, Israel, or even Qatar. Of course, this can make it all the more difficult and costly to litigate, so make sure you've considered the cost (in time, effort, and money) before moving forward.

Arbitration

In many legal contracts these days, you'll find a provision for binding or nonbinding arbitration. Basically, by signing the contract, you're agreeing that if you and the digital agency encounter a severe disagreement, you're willing to take your dispute to a professional arbitrator as an alternative to going to court. In the case of binding arbitration, you're basically *agreeing in advance* to abide by the arbiter's ruling, thereby waiving your right to sue. Both binding and nonbinding arbitration are acceptable alternatives; you should just be aware of whether your contract mentions this provision, and if so, be willing to agree to it.

◆ ◆ ◆

There's no way to anticipate everything that could go wrong with your website, but the more your contract can anticipate pitfalls upfront, the smoother your working relationship with your digital agency will be. If you make sure at least the basics described here are covered in the contract, you'll be in reasonably good shape.

PART THREE:
BUILDING A WEB PRESENCE THAT WORKS

CHAPTER 10

Budgeting For Your Website

So now we come to the topic that everyone loves to talk about—money! Planning a budget for a good website deserves its own chapter because your options are so diverse, not to mention the number of ways you can overspend. There really is no right or wrong answer as to how much you should pay (other than the typical "You get what you pay for" line). Mainly it's about how simple or complex your needs are: whether you need an original design or a template, how much content you expect to have, how much traffic you need to accommodate, and so on.

Essentially, there are three tiers of websites. Let's discuss each of them briefly.

TIER ONE: THE DO-IT-YOURSELF WEBSITE

If you've excluded your own time as a variable, the least expensive tier is the do-it-yourself (DIY) site. Providers such as Squarespace, Wix, GoDaddy's GoCentral, Shopify, and WordPress.com offer online services where you can quickly choose your own prefabricated template using their proprietary solutions. They can cost as low as six dollars and as high as five hundred dollars a year. While the DIY route is an inexpensive way to launch a website, there are some drawbacks to consider.

Software as a Service

All the providers in the DIY tier offer their solutions as a Software as a Service (SaaS). Essentially, your website belongs to another entity—it's a leased asset—and you can't transfer the site files to another provider should you choose to leave. If you select another SaaS provider, you'll need to redesign your site from the ground up using their software.

Limited options

Another drawback of the SaaS option is the lack of flexibility. Each provider will offer a preselected set of semicustomizable templates or themes from which to choose. Inevitably, there will be limitations within each template's framework, and you'll need to concede to those limitations if you want to use the template.

TIER TWO: THE TEMPLATE SITE

For a level up from the DIY site, you have the option to buy a predesigned, canned template to build your website. You can get these from one of the theme websites like Template Monster, ThemeForest, Array Themes, or CSS Igniter, and you can either have someone set it up for you or set it up yourself. There are editable areas within these templates, but the templates themselves retain their structure. You'll generally be able to change a few things as far as color or content, but by and large the original theme will remain. For this option, you might pay anywhere from $500 to $5,000, depending on the amount of customization you require. The good news is that once you've purchased the template, you can host it wherever you like, and you can retain ownership over the site if you change digital agencies.

With this option, you would hire a freelancer or digital agency to install the theme and the content management system. In all likelihood, the theme will require a fair amount of customization to reflect your branding, goals, and needs. While the installation of the theme may be inexpensive, the cost to customize it to suit your needs may reach into the thousands of dollars.

TIER THREE: HIRING AN AGENCY

The third tier, which we've been discussing throughout this book, is to hire a digital agency to design and build your site from the ground up. In this scenario, you express your wishes for your site, and the agency provides design concepts, branding, marketing, and so on. The only real limitation is your budget, because you're hiring a creative shop to customize the entire site to the needs of your business. The cost for this tier typically starts at about $5,000 and goes into millions of dollars, depending on how extensive your needs are.

A HYBRID APPROACH

If you want to use a template to build your site, but you lack the time or skills to customize it, some digital agencies (not all) are willing to set up your site using a template you have purchased.

The overall cost should be considerably less than a full custom project. If a fully customized Tier Three site isn't in your price range right now, this hybrid approach might be a viable alternative for you.

WHICH TIER DO YOU NEED?

There really is no right or wrong answer to this question. Rather, there are trade-offs associated with each tier. If you're going to go with Tier One or Tier Two, for example, you can't expect your site to be especially unique, because those templates are resold over and over again. You won't be able to accomplish a one-of-a-kind look because the framework is available to anyone who wants to buy it. So if uniqueness is a high priority, you'll need to find the budget to hire a digital agency.

On the other hand, if a quick release is your highest priority, you need to understand that Tier Three is probably not an option for you. If, for example, you have a time frame of a month or less—for example, you're pressed to get a blog or website up in time for your upcoming appearance on TV—most agencies won't be able to do custom development work that quickly. You'll likely have to use one of the DIY platforms or a template, at least for the short term.

WHAT IS THE WEBSITE WORTH TO YOU?

In evaluating all three tiers, you'll obviously want to ask yourself what a website is worth to you and your business. And if you're leaning toward the Tier Three option, this question is especially important.

We have a tendency to put a perceived value on the things we want or need, regardless of the actual cost. For example, some people are willing to pay $300,000 for a car; others will pay $30,000; and still others feel that $3,000 is too much to spend. There's obviously a huge difference between a $3,000 car and one that costs ten times that much, but the truth is, you can buy a car at any of these price points, and any number in between. It all depends on how much the car is worth to you, what you can afford to pay, and what you're *willing* to pay.

The same holds true with a website. You can build a website for cheap, or you can build one that costs you many thousands of dollars. Either way, you'll have a website. The important questions are how much website do you need, how much can you afford, and how much are you willing to spend? I urge you to spend some time figuring out your own perceived value for a website, and building a budget based on that value before you step into the process of choosing an agency. You don't want to start working with an agency if you're not clear about what you can budget.

KNOW YOUR NUMBER (AND DON'T BE AFRAID TO SAY IT OUT LOUD)

Once you've decided on your acceptable price range for a custom website, you should come up with a bottom line figure that is the basis for your budget. Then, when you're talking with a digital agency about doing the work, be up-front about your number. It doesn't mean the agency is going to spend all that money. Rather, it helps them frame the scope of work so it fits within your budget. So don't be afraid to say your number; it can actually be quite liberating. Nothing bad happens to you when you say it. In fact, it's the digital agency's job to work within that number and help you create a website within your budget. For example, if you say your budget is $10,000, your agency should be able to respond with something like, "Okay, let's see what we can do within that budget. The website might be completed at $7,500, and we can use the $2,500 on digital marketing."

TIMELINE

One of the components of budgeting that a lot of people overlook is time—in other words, how much time is needed to complete your project. For example, if you need your website up in time for a grand opening, a trade show, or something else that's time sensitive, you need to be prepared to pay a rush charge, or at least have an understanding of what can and can't be done by that deadline. Budgets necessarily increase when speed is a factor. If it would normally take eight to twelve weeks to do a website like yours, but you need it done in four, be prepared to pay a rush fee, or settle for less.

REQUEST FOR PROPOSAL AND REQUEST FOR QUOTE

Many companies, even small ones, will put together a Request for Proposal (RFP) or a Request for Quote (RFQ) that runs a prospective vendor through a series of questions. The purpose of this request is to create a uniform process for receiving bids for work. Government bodies typically use RFPs and RFQs to standardize their purchasing process. The website buyer decides on the basic elements they want in a website, develops an RFQ, and canvasses multiple digital agencies, asking them each to fill it out so they can effectively select the lowest bidder.

RFPs and RFQs are undesirable where web development is concerned, because these documents attempt to define, and by extension confine, what is a very nebulous product. Essentially, it's an attempt to compare apples to apples, and website development doesn't work that way at all. Website design is a creative process, and there's no gauge for creativity in an RFP. You're reducing a highly creative effort to a set of soulless parameters. Without a solid understanding of functionality, the users' needs, and the project goals, not to mention a strong relationship with your agency, the probability of failure goes up.

> **Website design is a creative process, and there's no gauge for creativity in an RFP.**

By approaching a digital agency with one of these documents, you risk starting off on the wrong foot. When you're hiring someone to be an agent for you in any capacity, you ideally find each other simpatico and are on the same wavelength. If you're sending out RFPs or RFQs, you're basically approaching a digital agency with whom you have no creative relationship and asking them to take a test. This *might* work if you're using the hybrid approach—that is, if you've already selected a service or template and you just need someone with some technical ability to fill in the blanks for you. If that's all you're looking for, by all means send out an RFP. But if you're looking for a relationship with someone who understands you and serves as your digital agent, an RFP is simply not the best way to go about it. In fact, many of the best digital agencies who do the best work won't even respond to an RFP because (1) it's a bid check, and (2) if that customer views the agency's work as nothing more than a commodity, why would they want to do business with them?

As the head of a busy, successful digital agency, I'm looking for likeminded new clients—people who value my work. I'm looking for people who send me an email that says, "Drew, I saw the website you did for so-and-so, and it is fantastic. I'd love to talk to you about doing a website for my company." That's entirely different than sending an email that says, "Here's our RFP, please reply by June 15." If you were me, which would you rather answer?

There's this dynamic in the marketplace that I like to call the buyer-seller dance—you have a buyer with a need, and a seller who can meet that need. The two parties dance around the details and the budget until they arrive at a solution that is tailored to the needs and budgetary requirements of both. Like any dance, it's a bit awkward until you learn the steps, but with creative endeavors like website development, it's really the best way to do business, because it builds a high level of trust between the buyer and seller. The RFP effectively shuts down the buyer-seller dance, which is why most worthwhile digital agencies won't even touch it. In fact, I dare say if you send an RFP to a digital agency and they respond to it, that in itself could be a red flag.

The bottom line is that budgetary needs aside, you need to form a relationship with your digital agency. Unless you take the time to create that relationship, you're missing out on some creatively talented people who could otherwise do fantastic work for you. If you have a custom job in mind and you're looking for the best and brightest, go meet them. Pick up the phone and call them. Otherwise, you can expect a website that is boring at best, and dysfunctional at worst.

CHAPTER 11

Building and Developing Your Website

Now begins the exciting part. After all the effort you've put into choosing a digital agency, working out contract details, and figuring out your budgetary needs, it's time to begin working with your agency to bring your website to life.

Building a website is both a collaborative and creative process. Depending on the extent of your project, you'll be working closely with your agency over the next few weeks or months in a back-and-forth process as the various steps of planning, design, and development take place.

If you recall from Chapter 2, we discussed the difference between web designers and web developers. To recap, *web design* refers to the front end of your website. Essentially, it's the overall look and feel of color schemes, graphics, layouts, and everything else that goes into the user experience. *Web development*, on the other hand, has to do with back-end coding. Your users will never see this part of your website, but it serves as the backbone for how your website functions.

Think of design as the dining area of a restaurant, while development represents the kitchen. Customers aren't bothered with what the chefs are doing in back—they just come to enjoy a fine meal in a relaxing environment. Similarly, development and design work together to provide a satisfying user experience.

During your website project, it's your job to provide designers with input regarding how the site should look and feel. Once those decisions are made, the web developers will work their magic, writing the code to bring those designs to life.

THE STAGES OF BUILDING A WEBSITE

A reputable digital agency will have a tried-and-true step-by-step process to build and launch your website. While you don't need to know all the details that go into each step, it helps to have a basic understanding of the process, so you can not only communicate more effectively with your team but also easily track the progress of your website. At several points along the way, you'll be asked to review certain features of the site to provide approval or recommend changes. The order of the steps listed below might vary slightly from agency to agency, but the elements are generally the same for any website project.

Discovery

During this first stage, your digital agency begins gathering all the information they will need to carry out and complete your project. They will likely ask you to discuss the nature of your business, your goals, and your desired audience, as well as the features and functionality you want. From these conversations, the agency will begin formulating a plan of action.

Asset gathering

Once the agency has a good idea of your needs, they will begin assembling a set of *assets*, or elements that will either inform your website or become part of its content. These assets may include graphics that you provide (including your logo—more about this topic later), graphics acquired from elsewhere (stock photos, affiliate logos, etc.), background information on your company, any existing relevant written content (mission statements, team bios, etc.), any social media sites you plan to link to your website, and so on.

Content writing

While the written content of your website will be created later in the process, your digital agency will probably begin discussing this step with you early in the project because developing the content may take some time. You may opt to provide your own written content, or if you don't have someone on your team with the skills or the time to write it, your agency may arrange to produce the content for you, with your input. Either way, your written content should have the following characteristics:

- Original to your site (i.e., it shouldn't appear anywhere else online)
- Free of spelling and grammar errors
- Accessible to a broad audience
- Appropriate voice and style

That last point is particularly important. Your content should be written in a professional voice appropriate to your brand and type of company. After all, content for a law office won't read the same as content for a beauty ecommerce store.

And don't cut corners during this step—your written content is every bit as important as the way your website looks.

Wireframes

Wireframes are like a skeletal structure for your website design; they're a visual blueprint or mock-up with placeholders, so you can envision where various features will go and how your website will look. You'll be asked to approve this blueprint or give your input for changes.

Design

At this point, it's time to create the actual design of your website, from color scheme to layout to content placement. Feel free to offer input, especially when asked, but this is the point at which you'll want to give your digital agency some creative liberty—they might surprise you! Once they've done an initial design, you'll go through one or more rounds of edits to hone and refine the design until you're happy with it.

Development

Once you've approved the website design, the developers go to work writing the code to format your designs for the web.

Quality assurance and testing

Once your website is coded, it should go through an extensive quality assurance (QA) testing process to make sure all the elements work as they should. Do the colors and layout look correct? Do all the links and buttons work? Do the forms get sent to the correct email address? Depending on the scope and complexity of your website, it may have to go through several rounds of tests, even after launch, to make sure all the bugs are worked out.

Website launch

Congratulations! Now you have a live working website. Go through and check everything. Click all the buttons and links. Make sure the site looks and functions as it should. Then begin spreading the word to your customers.

Maintenance

The maintenance stage of your website is typically ongoing, depending on your arrangement with your agency. The agency will perform periodic tests and updates to make sure your website stays functional as various web protocols are continually improved. Servers may be monitored, upgraded, or changed out to prevent unwanted downtime.

Promotion

Many digital agencies also offer marketing and promotion services designed to help increase traffic to your site and grow your customer base. We'll talk more about these services in the chapters ahead.

ESTABLISHING A TIMELINE AND MILESTONES

It's often said that a goal is a dream with a deadline; this is exactly how you should approach building a website. Every website project should have a definitive timeline assigned to it, along with a set of agreed-upon milestones at various points on the timeline. Think of milestones as stops along the journey of putting a website together. Just as you might plan a cross-country trip where you'll arrive in certain cities by certain times, your website project should have defined markers with projected times of arrival. When will the logo be ready? When will the content be ready? When will the wireframes be done? When will the design be complete? When will the site launch? All these milestones—and any others agreed upon by you and your digital agency—should have definitive dates on the calendar, so everyone can stay on track.

> **Every day your website is not online is a day that you're not capturing leads.**

Why are deadlines and timelines so important? If there's no project timeline in place, the website building process could go on indefinitely. And every day your website is not online is a day that you're not capturing leads, making online sales, or putting your best foot forward with your clients.

Once your project begins, a good digital agency will usually put together a timeline for you, with milestones you can agree upon. If the agency doesn't initiate this process, make sure to ask for it yourself. It's also a good idea to overlay your calendar onto the agency's calendar so you can identify any potential timing conflicts. Be clear about when you will and will not be available. For example, if you know you're going on a beach vacation for a week, you should probably let your agency know about it and not set any due dates for that week. If you have another project at work that won't allow you to give adequate attention to the web-development project, ask the agency to build some extra time into the project schedule. And if your website needs to be up and running in advance of a major business event like a trade show or a new product launch, tell your agency in advance so they can plan around it.

COMMUNICATING DURING THE PROJECT

You'll want to establish regular and consistent points of communication with your digital agency over the course of your project. Depending on the scope of the project and the firm itself, it's common practice to set up a standing phone call (usually weekly or even daily) so you can receive regular status updates, share your feedback, and discuss what's needed next. Set these touchpoints as firm dates on your calendar so you'll know when to expect the next call (and so you don't have to keep calling your project manager to check on how things are going). A standing call also helps ensure that you have the agency's undivided attention, as opposed to calling when they are possibly distracted by other tasks.

In case your agency needs to get ahold of you between your regular calls, let them know your preferred communication style. We mentioned this earlier, in Chapter 2. Some people prefer email or text, while others prefer talking on the phone. Usually a digital agency will be fine with any of these, so just be up-front about your preferences.

Establish a point of contact

To further ensure good communication during your project, assign upfront a project leader or spokesperson from your company who will interface with the digital agency. At our firm, we prefer one person to serve as the liaison from the client company. That person is responsible for gathering any and all feedback from other decision-makers or stakeholders in the company and relaying it back to us. That way, institutional knowledge is preserved with that one person, and we avoid getting cross talk, with six or more people copied on an email thread (and a "too many cooks in the kitchen" situation). Interoffice communication can exist within your company however it works for you, and of course you can have as many people as you like within your company contributing to the decisions. But when interacting with the agency, try to funnel the communication through one designated person who speaks for everyone at your organization. Give your organization one consistent voice.

Consider a chat or instant messaging program

If you and your digital agency prefer more frequent or instant communication for greater connectivity during the project, you might want to consider connecting via one of several chat or instant messaging apps. In our firm, we use Slack because of its collaborative features, which enable chat and file sharing in the same app. Be advised, however, that there are pros and cons to using this kind of service. While it provides more instant communication between agent and client, if too many of your teammates are involved, it can also generate the same kind of cross talk that's fostered by email threads, as discussed earlier. Ultimately, it's a matter of preference. An app like Slack can be a useful tool in collaborating on a website project, but if you use it, you might want to set some ground rules with your teammates to keep communication from getting too busy and cluttered. Maintaining one point of contact with the agency can definitely help.

YOUR RESPONSIBILITY TO THE PROJECT

Throughout the course of your project, your digital agency is likely to assign you certain tasks that you'll need to complete by the agreed-upon deadlines to keep the project on track. If you're generating the written content for the site yourself, you'll need to specify who is delivering the content and by what date, so there's some accountability in place. If you've been asked to deliver other assets such as graphics, social media links, or your logo, it helps to make a checklist of these items so you can cross them off as you deliver them. Being organized on your end is as important as it is on the agency's end.

Providing your company logo

If you're like most businesses, your company logo is an essential part of your brand. As such, it will also be an essential part of your website. If you don't have a logo yet, you'll want to have one developed, whether you contract it, crowdsource it, or have it created by one of your own employees. If you already have a company logo, but it only exists in low resolution, you'll need to have it remade into vector format, which is a high-resolution format that your digital agency can scale to any size. When it comes to graphics, you can always go *down* in quality, but you can't often go *up*. In other words, you can always shrink a crisp high-resolution logo to fit a certain space, but making a low-resolution logo larger can make it look blurry and pixelated. A vector logo works in any setting.

As an additional note of caution: always keep a copy of your logo in your own files. You'd be surprised how often I ask a client for a copy of their company logo, and they tell me, "Let me get that from my sign guy," or "Let me ask my previous web designer for a copy of it." This always troubles me, not just because it creates delays in the website project, but because it represents a potentially massive point of failure for one of the biggest assets of the company brand. Anytime you develop a logo, always keep a high-resolution copy of it in a safe place *in your own files*, preferably backed up. Not only does this make things much simpler for your digital agency, but it protects you from a lot of headaches down the road.

◆ ◆ ◆

Once your project is complete, launching your new website can be one of the most satisfying moments for your company. Now it's time to make the most of that online presence. In the last two chapters, we'll talk about some online marketing concepts that will help you do just that.

CHAPTER 12

The Ins and Outs of Digital Marketing

So your website is finally up and running. Now what? Will emails from prospective customers start flooding your inbox? Will your ecommerce sales go through the roof?

Not likely. Not yet, anyway.

It's important to understand that having a website and driving traffic to that website are two entirely different things. For many business owners (not all, as we'll see shortly), developing and implementing a solid digital marketing strategy is the logical next step after launching a high-quality website. That's why many digital agencies will offer digital marketing services in addition to building and running your site.

DO YOU NEED DIGITAL MARKETING?

The first thing to ask yourself is whether you actually *need* to market your website, because not all businesses do. Depending on the purpose of the website and the type of business you run, some companies will do fine with just a web presence and no additional digital marketing.

If the primary reason you have a website is to legitimize your company for potential customers who search your name, then you probably don't need much digital marketing. Displaying your web address on business cards, letterhead, or brochures may be the only marketing you need, since you don't really conduct your business online.

Likewise, if your business is done strictly by referrals, you probably don't need to market, because you're likely to get the wrong phone calls if you do. For example, one of our clients is an anesthesiologist. No one sees him without a referral from a primary care physician. So he doesn't need a marketing plan for his website, but he *does* need a website, because it legitimizes his practice in case someone wants to look him up before going under the knife. The website is worth it, even if it mainly serves to make his patients feel better about him. But if someone were to find him from an online search and call his office, it wouldn't be helpful to either the doctor or the patient.

However, if you plan to use your website to either generate leads or sell your products or services online, a solid digital marketing strategy is not only beneficial, it's essential. If you've never carried out an online marketing effort before, it's important to realize that success hinges on a slightly different approach than other types of marketing. It requires you to define your audience and take specific steps to reach those people—and *only* those people—online. Let's dig a little deeper into this.

THE BROAD AND THE SPECIFIC

Most marketing strategies can be defined by one of two approaches: broad and targeted.

Most forms of mass advertising rely on the *broad approach*, in which you're aiming at a wide audience and hoping to make some impact on a large population. Examples of the broad approach include advertising on TV, billboards, or buses. You're paying for a mass number of eyeballs to see your ad, but not necessarily the right eyeballs—that is, quantity, not quality. You're only loosely targeting your audience by placing your ad where your customers *might* see it and remember the name of your business. The broad approach can be expensive, and you have to do it consistently and for a long time before you begin building your brand recognition and seeing results. When it comes to digital marketing, the broad approach isn't recommended unless you have a long timeline and a large budget.

The other approach, and the one that tends to be more effective with digital marketing, is the *targeted approach*, in which you strategically narrow your aim to a specific target audience. The idea is to help your website show up near the top of online search results so people can find you more easily. This practice, known as search engine optimization (SEO), is typically accomplished by associating your website with search terms and keywords that your customers are most likely to use when searching for your type of business.

I typically advise my clients to take this more strategic targeted approach to digital marketing because (a) it will cost them far less than mass advertising, and (b) the people they reach with this method are more likely to be within their target audience. To start, I advise my clients to think of search terms that accurately describe their business. These search terms can then be refined into a set of keywords or keyword phrases. From there, they can choose either an organic or a paid approach to search engine results, or a combination of both.

ORGANIC SEARCH RESULTS

When going the organic route to keyword-based marketing, you craft your website content around your chosen keywords with your goals in mind. Search engines organically index website content based on *relevance* to the search term. In other words, when someone searches for "fashion jewelry in Miami," the search engine will recommend websites featuring content that fits that description. This organic method of digital marketing can cost nothing but time if you write the content yourself; however, it also requires patience because it may take four days to four weeks before your site begins to show up in the search results. Why? While the search engines constantly scour the web for new information, they often do not immediately include new sites and new pages in search results. Factors like site maturity, content value, and relevance all play parts in search engines' wildly complex algorithms.

Additionally, making your content keyword relevant (or "SEO friendly") can be a guessing game if you don't have a well-constructed content strategy. If the prospect of spending hours learning how to develop title and meta tags, creating rich content for your site that drives traffic, and building links from other websites to yours don't sound appealing, you might consider hiring an expert to help you.

PAID SEARCH RESULTS

With the pay-per-click (PPC) or paid route to search results, you're basically paying Google, Bing, and other search engines for higher levels of visibility in their search results based on a list of your chosen keywords or keyword phrases. Searches using your keywords will generate an ad on the engine's pages of search results, and you'll pay a certain amount for every visitor who clicks on that ad, which links to your website. You pay for the click (or *clickthrough*), but the amount you pay depends on the competitiveness of the keyword. In most cases, the amount varies based on how general or specific the search keyword is.

When determining your list of paid keywords, it might help to visualize a Christmas tree, starting narrow at the top and widening nearer the base. At the top of the tree will be your broad, single keywords; you will pay the most for users who come to your website using these search terms. Under that would be two-word keyword phrases, which may cost a little bit less, then three-word and four-word keyword phrases, which cost even less. The rule of thumb (with a few exceptions, of course) is that the more specific you can get with your keywords (especially the longer phrases, or *longtail keywords*), the less you're going to pay for those keywords. Also, the more specific you get, the more targeted you're going to be in reaching your ideal audience.

When you're compiling a list of possible keywords, try to limit yourself to a handful. Focus on those keywords for now, and as you start to perform really well with those, you can consider adding to the list. People commonly want their website to show up for every conceivable search term, believing that will draw in more traffic. However, a broad strategy will cost much more than a specific one. Avoid mass advertising—you're looking for the right sets of eyeballs to view your site. Your digital agency may be able to advise you on the best keywords to use for your type of business.

GEOTARGETING

When coming up with keywords, you want to keep in mind where your customers live. Do they live in your neighborhood, or in a wider area? Are your customers local, national, or worldwide? For businesses trying to reach a local or regional audience, it's very helpful to come up with key terms that include location.

This strategy is called *geotargeting*, and it can be a highly effective approach to keyword-based marketing, because you're able to reach your local audience without having to compete regionally or nationally with other businesses' general search terms like "best pizza" or "coffee shops." With geotargeting, you're likely to pay less and reach your local audience as well.

Geotargeted keywords can and should be very specific, especially when you're trying to reach a specific neighborhood. For example, if I live in Brooklyn, New York, and I'm looking for a pizza joint, the keyword phrase "pizza shops in Brooklyn" will actually yield far too many results to do me any good. After all, Brooklyn encompasses more than four million people living in more than sixty neighborhoods, and I'm probably not going to want to walk more than five blocks to get a slice of pizza. As a hungry customer, I'm more likely to use a search term that includes my neighborhood, like "pizza in Park Slope." If you happen to own a pizza shop in Park Slope and you've used that geotargeted keyword string with your website—boom. I found you. I'm on my way to get a slice.

Geotargeted keywords work with just about any type of local business. If you're a CPA in Washington Heights in Manhattan, you could do very well with the keywords "Washington Heights CPA." You could also do well with "Manhattan CPA," but you'll have a bit more competition. If you use the keywords "New York CPA," you're now covering a whole state, which isn't very effective because a family in Albany isn't likely to come down to Manhattan to do their taxes.

COST PER ACQUISITION

When you're developing a digital marketing plan, it's very important to know your *cost per acquisition (CPA)*—that is, how much it costs you to obtain each new customer in your type of business, or how much it *should* cost.

While it's not actually presented this way, Google AdWords is an auction in which the highest search results go to the highest bidder. So in the mortgage industry, for example, the fat wallets at the table are likely to bid higher in the auction. As a result, "mortgage" is a keyword with a very high cost per click.

On the other hand, let's say you run a niche business like an Italian ice stand. If you're choosing three- and four-word keyword phrases that target only a small area, like "Italian ice stands in Park Slope," you may pay pennies for each clickthrough. And yet, those cheap clickthroughs are going to be high-quality leads because those customers are specifically looking for your business. This is the perfect example of a targeted approach.

The targeted approach to digital marketing works particularly well for smaller businesses who can't afford the high cost of branding through mass advertising. If you're an individual mortgage broker, and you're trying to compete against a national mortgage lender for visibility in search results, let's face it—you can't. So you have to find alternative types of marketing to make your business stand out, whether it's a Facebook ad, a newsletter, or specific search engine keyword phrases that target potential customers in your neighborhood. All are potentially effective and affordable ways to market your company online without trying to compete with national companies who have seemingly unlimited advertising budgets.

THE PROBLEM WITH CONVERSIONS

An often misinterpreted word in the digital marketing space is the word *conversion*. It can mean different things to different people. For some digital marketers, a conversion occurs when someone clicks on a link on your website. Other marketers consider user engagement a conversion only if someone signs up for more information. And others consider a conversion to be when the prospect makes a purchase on the website. It's very important to be clear with the people on your team and those assisting you on what your interpretation of the word *conversion* is. Having a shared understanding of what it means helps ensure that everyone working on your digital marketing project is using the same language.

POTENTIAL DANGERS WITH PAID ADVERTISING

Although digital marketing can be quite affordable, you still need to be careful not to overspend, especially when it comes to pay-per-click advertisements, like Google AdWords, Bing Ads, or Facebook Ads.

Typically, I believe in the 1:100 ratio when it comes to marketing: you have to reach one hundred targeted people with your marketing in order to get one person to take some action. And that action isn't necessarily buying your product or service, but perhaps just inquiring about it—filling out an online form, calling for more information, or downloading a white paper. Then you have to figure out how many of those people you have to include in your conversion funnel—or sales path—to make one sale. For example, if only one out of five people who reach out to you or take some action on your site (e.g., download a white paper) eventually becomes a customer, then you are going to need to reach five hundred people, at the top of your funnel, to generate one sale at the bottom end of the funnel.

To further illustrate this, let's say you have $1,000 to spend on paid advertising. While that might sound like a large budget, it may not be nearly enough. For example, if it costs ten dollars per clickthrough, you can safely expect roughly a hundred people to visit your site. So if only one out of a hundred people who visit your site calls you, completes a contact form, or orders your product, you should expect only one sale for your $1,000 budget.

If fewer than one out of a hundred people engage with you after visiting your site, you may not receive any leads that month. And if your cost per click is greater than ten dollars, the number of visitors who come to your website from the search engine ads will diminish.

Yes, it's very possible that after spending $1,000, you may not get any leads at all.

You might get a nibble with paid ads the first time you try. Perhaps you'll get a call after a couple of days. It could be another sixty days before the next call happens, but by then you will have put more money into paid ads because you thought they were working. This is exactly why you need to know your cost per acquisition, and why you need to home in on the specific, targeted keywords that work for your business.

Another way to protect yourself when paying for clickthroughs is to know your margins. Let's go back to the mortgage broker example. Let's say a mortgage broker earns $5,000 with the average mortgage, and he needs a hundred clickthroughs to get one conversion. In that scenario, at fifty dollars per clickthrough, everything the broker makes on that one sale is going back to the search engine, which obviously isn't cost effective. That broker needs to decide how much of his commission he's willing to spend on marketing, whether that's a 5 percent or 15 percent margin. In other types of businesses, an acceptable margin could be much higher. Some venture capital–backed businesses will spend far more than their income in an effort to market their brand, hoping it will catch fire. Google AdWords can certainly be an effective way to help you find customers, but you need to decide for yourself what your acceptable margin is before putting the money out.

DETERMINING HOW MUCH A PAID KEYWORD WILL COST YOU

If different keywords cost more per clickthrough than others, how do you know your cost per keyword? Fortunately, most of the paid advertising players have tools where you can type in any keyword or keywords and see what it costs to bid effectively on that term. Bear in mind that this will only be an estimate, because the cost displayed is an average of sorts. You can actually drive your cost down by being the best match for the keyword you're using.

For example, if the keyword you're buying is a precise match for the ad text you're using and links to a page on your site that only talks about the keyword, you're likely going to pay far less than an advertiser who has generic ad text that lands on their company's homepage without any mention of that keyword.

Let's return to the example of the Italian ice vendor in Park Slope. You are that vendor, and Italian ice is literally all you sell. You use that keyword phrase: "Park Slope Italian ice." Your Google ad says, "Looking for Park Slope Italian ice? Click here for the best in the borough." When someone clicks on that ad and lands on your website's homepage, all that page talks about is Italian ice in Park Slope—no custard, no ice cream, no gelato. To the search engine, that's a perfect 1:1 match between the search term and the intended result. For that type of perfect match, you'll pay far less per clickthrough than if you had paid for the terms "summertime novelty," or "desserts in New York."

SOCIAL MEDIA MARKETING

I won't attempt to go into all the detailed strategies around using social media (e.g., Facebook, Twitter, blogs, etc.) to market your business—that's a book for someone else to write. But purely from the perspective of effective marketing, if you're wondering whether you should use social media as a tool, here's a bit of common-sense advice.

> **Just because the other kids are doing it doesn't mean you need to do it, too.**

Do you really need social media?

It's the same question you should ask yourself regarding digital marketing. My anesthesiologist client doesn't need social media for the same reasons he doesn't need to market his website—his business comes from referrals. In fact, if he attempted to use social media, it might actually have the opposite of the intended effect and delegitimize his practice in the eyes of some. Why? Because it isn't a good match for the type of practice he runs. So make sure you need social media before spending your time on it. Just because the other kids are doing it doesn't mean you need to do it, too.

Finding your audience

If you determine that social media is a good way to reach your intended audience, your next step is to decide which of the many social media platforms you should join. There is a superabundance of them out there—Facebook, Twitter, Snapchat, Instagram, Pinterest, and so on. It's very easy to get lost in all of it, and there's no way you're going to do all of them well—keeping up could easily become a full-time job.

Remember: *you run a business.* That's what you do well. You're a great writer, or a great veterinarian, or a great chef. You run a great bed-and-breakfast, or a great Italian ice stand. You're not the queen of Snapchat. So my advice is to focus on one social media outlet—or two at the most, at least to start—that best fits the nature of your business and your audience.

Choose a channel that you know your audience uses. If my audience is under twenty, and I happen to like Snapchat, that's probably going to be a great connection point. If I run an Etsy store, and I'm using Pinterest to post pictures of my macramé—boom. But if my audience is looking for retirement communities, and I'm posting videos on Snapchat, it's pretty much guaranteed I'm not reaching them.

Authenticity rules

I'm serious on this point. If you're going to market via social media, it's usually best to do it yourself or have it done by someone who works at your organization. Personally, I'd never hire an outside firm to run my social media accounts. The reason is simple: half the term *social media* is the word *social*. How is someone who isn't an employee or steeped in your company's culture supposed to reflect that culture? How can an independent contractor who doesn't live and breathe your ethos communicate it to the outside world?

Consumers can smell a lack of authenticity a mile away.

One of the intentions of this book is to help you be authentic as a businessperson—with your budget, with your timeline, and with your online presence. How can you hire someone to be your surrogate—someone who isn't you or someone close to you—to be chatty online with your current or prospective customers? Eventually, it's likely to go sideways. You might be tempted to hire some college student because she's good on Instagram, but that college student does Instagram because it's a reflection of who *she* is and what *she* likes, not necessarily who *you* are, either as a person or a company. So find a social media outlet that reflects you and power it internally. Hiring a third party to do it is inauthentic, and social media consumers can smell a lack of authenticity a mile away.

WHAT ABOUT BANNER ADS?

Online banner ads, also known as *display ads*, seem like an attractive way to gain new customers at first, but they follow a broad approach, which means they aren't particularly effective when compared to other specific marketing tactics. I rarely recommend them because the clickthrough rates (CTRs) are abysmal. (Remember the 1:100 ratio I talked about with regard to search advertising? Banner ad clickthrough rates are actually much, much worse.)

The only reason to consider buying banner ads on the Internet is for the purpose of brand recognition, because it gives you a visual canvas. For example, if you're running for office, and as a voter I see your banner ad telling me to vote for you, I'm probably not going to click on it, because I know it's going to take me to a donation page. However, seeing that ad does create an impression that I'm more likely to remember later, namely when it's time to vote. Banner ads are good for long-term impressions and brand recognition, but not for direct sales.

• • •

As you can probably tell by now, digital marketing is something of an art form all its own. There are even more nuanced details to it outside of the brief overview we've given here, which is why if you're serious about establishing a successful online presence, you might want to consider hiring your digital agency to help you implement a digital marketing strategy customized to your type of business. Before we wrap up, though, there's one form of online marketing that I personally believe is both underrated and underutilized, but which can be highly effective if implemented correctly. We'll talk more about it in the next chapter.

CHAPTER 13

Email Marketing

There's one facet of digital marketing that can be highly effective, but it's also so simple that it's often overlooked. It's not very flashy or technologically impressive compared to other features of the Internet, but when leveraged correctly, it can accomplish more than nearly any other online marketing technique. That's why we're about to spend a whole chapter talking about it. I'm speaking, of course, about the wonderful world of email marketing.

It seems that nearly everyone these days has an email address, and perhaps it's precisely *because* email is so commonplace that we underestimate its potential as a marketing tool. But we shouldn't.

An email address is a powerful thing. It is a direct line of communication with potential clients that is more instant than a physical "snail mail" address. What's more, anyone who consents to share an email address with your company is essentially inviting you to use it. That means the odds of converting an email subscriber into a customer are far greater than someone who simply stumbles upon your company website.

And yet, for all this potential, many companies don't use this tool effectively. Far too often, companies are lax in obtaining email addresses from their existing clients to keep them abreast of changes, updates, specials, or sales. Even when they *do* compile an email list, they send emails either too infrequently or too often. However, when a company finds that happy medium between regular email communication and offensive spam, email marketing can become one of the more lucrative aspects of the business. That's why some of the best digital agencies include email marketing among their product offerings for their clients.

There are entire books written about email marketing, so if you're interested in delving deeper into the subject, consult them or the online tutorials that are available to you. The tips and advice below should give you an overview on how an effective email marketing strategy can enhance your web presence, whether you do the marketing yourself or hire a digital agency to manage it for you.

BUILDING A QUALIFIED LIST

What's so advantageous about building and managing an email list? For one thing, when you build your list correctly, the people on it are all fairly good leads, because they've already expressed an interest in your business and are therefore more inclined to use your product or service. For another, email marketing gives you the ability to tailor your message to your audience and say exactly what you want to say, even to the point of sending specific messages to certain subgroups within your list. And perhaps most importantly, email lists offer several key metrics that enable you to see exactly how your recipients are interacting with your emails. This gives you the opportunity to further refine your message.

That said, one of the keys to a successful email marketing campaign is to build a *qualified* list, meaning that all the addresses are from people who shared their email with you for the purpose of receiving communications. If you have a sign-up on your website for people to join your email list, this permission is implied—they wouldn't be signing up if they didn't want to hear from you. If you obtain email addresses from people by giving them a paper form to complete (for example, in person at trade shows), it's best to *explicitly* state that they are giving you permission to communicate with them via email. Anything less than this, and you'll likely be spamming them.

An email subscriber is trusting you with access to a personal asset, and misusing that asset represents a breach of that trust.

Whether you obtain email addresses in person or online, you should be as clear as possible in disclosing how you intend to use that person's email, how you're protecting their privacy, what kinds of email you'll be sending, and approximately how often. When you ask for someone's inbox real estate, there's an understanding of trust that you must respect. That subscriber is trusting you with access to a personal asset, and misusing that asset represents a breach of that trust.

When building your list, remember to prioritize quality over quantity. The objective isn't just to collect as many email addresses as possible, but rather to connect with as many *qualified leads* as possible. This isn't just to avoid spamming them—it's actually in your best interest to have people's permission to email them. When your list only consists of people who have expressed an interest in your business, you'll be in a better position to turn most of those subscribers into eventual customers.

TAILOR YOUR MESSAGE WITH ENGAGING, TARGETED CONTENT

If you have an email newsletter, the best way to keep your subscribers happy is to give them engaging content. If the content isn't any good, or if it's inappropriate and unappealing, your readers will delete the emails without opening or reading them—or worse, they will unsubscribe. The content needs to be enticing, valuable, and relevant to the reader.

Of course, it's difficult to create content that is relevant to everyone on your list. Your readership is probably fairly diverse, and not everyone is going to relate to everything. One of the best ways to address this issue is to segment your list.

For example, let's say you're opening a new store in Atlanta. It doesn't make any sense to send your grand-opening announcement to your subscribers in New York, because those people aren't going to fly to Atlanta for your store opening. Your customer in New York would see the announcement and wonder, "Why should I give a damn?" The more you send "Why should I give a damn?" stuff to that customer's inbox, the sooner the customer *won't* give a damn and will unsubscribe.

The solution to this dilemma is to know your audience and segment your list accordingly, whether by location or by the types of services you provide your clients. For example, if you're an event planner, it makes no sense to send wedding planning content to your corporate events clients. So don't; segment your list according to interest so the wedding planning content only goes to your clients who might be interested in it. This way, the content you send your different subscribers is relevant and engaging only to them, and they're more likely to keep reading.

THE ALL-IMPORTANT METRICS (ESPECIALLY THE BOUNCE LIST)

One of the best things about email marketing is that you can see what people do with the emails when you send them. You can see who opened your email (and who didn't), what links they clicked on within the email, and how many times they opened the email, among other things.

These metrics are incredibly important for refining an email campaign. With metrics, you can do some *A/B testing*, or see how changing different aspects of your email impacts the metrics. For example, if your initial open rate is low, some email newsletter software will allow you to experiment with different catchy subject lines to see if they improve your open rates; or you can see if sending a biweekly email gets you more results than sending one monthly.

Within these metrics, one of the most important numbers you should be looking at is your *bounce list*—all the emails that were returned to you because the recipient email addresses are no longer active. If you're building a list, a bounced email is the first indicator that something's gone wrong, or that there is a change you need to heed. If a customer's email address bounces back—and it's a work email—it typically means that person is not at the company anymore, which means that the company won't be ordering from you unless you can reach your original contact's replacement. It's a lot easier to keep your existing clients than to replace them; your bounce list gives you the opportunity to take steps to reconnect with and retain your clients.

If you're in a service business, there's no barometer more important than your bounce list. For example, let's say you're a real estate agent, and you notice that the email address of a past client is no longer active. Again, if it's their work email address, that probably means they're not at that company anymore. When you're a real estate agent, your business often revolves around your clients' life changes—when they get a promotion, lose their job, get married or divorced, change neighborhoods, etc. If life change is where you make money, your bounce list is invaluable. Paying attention to this metric can provide you tangible opportunities to make sales as you reconnect with those clients.

HOW OFTEN SHOULD YOU SEND EMAILS?

As we said earlier, some companies send emails too frequently, and others not frequently enough. How do you find that happy medium between those extremes? How often should you send emails to your list?

The answer will differ according to several factors, and you may have to test things out and make adjustments accordingly. You'll want to look at your metrics along the way to evaluate whether your emails are effective. In particular, you need to follow two big indicators: your *unsubscribe rate* and your *conversion rate*.

Your unsubscribe rate

You need to look at how many people are unsubscribing from your list, and at what rate it's happening. If you're getting an unsatisfactory number of unsubscribes, typically higher than one percent, it's a strong indicator that your readers aren't happy with the content and/or the frequency of your messages.

Your conversion rate

Your conversion rate—in the context of email marketing, the number of email recipients who actually end up buying from you as a result of your emails—is the metric that helps you determine whether your campaign is actually working. Are you getting orders from your emails, and if so, how many on average? You can also test to discover the point at which you see a diminishing rate of returns. Perhaps you start with emails once a month, and that goes well; so you adjust it to once every three weeks, and then every two weeks, both with similar positive results. But when you start emailing every week, you start seeing your conversion rate drop. This is an indicator that you might want to back off a bit and return to biweekly. This type of testing helps you gain optimal traction and improve your conversions.

HOW LONG SHOULD YOUR NEWSLETTER BE?

There are several schools of thought on this question, but I have a philosophy that you can't keep your newsletter short enough. When you're composing email newsletters for your subscribers, your primary goal should be to move the reader from their email client to your website in as little time as possible. Sometimes, this can be accomplished with as little as one call to action and a link within the newsletter. Other times, depending on your product or service, you might want to include a brief article or story to engage the reader's interest before the call to action. Imagine that the reader on the other end has a short attention span, and your goal is to spark just enough interest to motivate the reader to shift from scanning their inbox to shopping on your website.

DON'T OVERLOOK THE "REPLY" BUTTON

In an email marketing campaign, a link to click through to your website isn't the only call to action. Another subtler call to action that shouldn't be overlooked is the "Reply" button. In many cases, the client will use that button as a method to communicate with you. Often, the subscriber's reply will have nothing to do with what was in the body of the email itself, but it does reflect what is important to that customer at that time. If you're paying attention, this often translates to sales.

The reason some subscribers will hit the "Reply" button is because they don't make a distinction between email marketing and direct email communication with them; if they get an email from you, they assume it's from you, and that there's someone on the other side who will listen to their needs. If you recognize and respect this, it can really work in your favor. For this reason, I typically advise my clients to avoid using no-reply email addresses, because someone is going to want to communicate back with them at some point. Even if the reply is an unsubscribe request, it's still a response to a call to action, and that's data you can use to hone your approach.

TIPS FOR WORKING WITH AN AGENCY ON AN EMAIL MARKETING CAMPAIGN

In the previous chapter, I recommended keeping social media marketing in-house. Email marketing, however, is something that can be outsourced, and if it's not something you can keep up with on a consistent basis, you *should* outsource it to a digital agency rather than let it go by the wayside. If you don't have a marketing team that can create and send regularly scheduled emails, you need to employ the services of someone who can. If you set up the workflow properly, you'll be able to review everything in the newsletter before it goes out, so it will sound like your voice.

Before you hire an agency or third party to do your email marketing, you'll want to see some of their previous work to make sure you like the quality of work they've done in the past. You'll also want to take a look at their conversion rates, if they'll share that data with you. How many people clicked on their links, and how many people purchased a product or service as a result of their emails? Also, in your service agreement, be sure the terms of your marketing campaign are as specific as possible, especially when it comes to scheduling. It's not enough for the agreement to say, "We're going to send out one email campaign a month." That language isn't sufficient. It needs to say something more like, "We will provide a draft by this date, and it will launch on this date." It should also account for certain times when it might not be beneficial to send an email. For example, if the monthly release date happens to fall on a Saturday, when fewer customers will be reached, what day will the agency use as an alternative?

CONNECTION LEADS TO CUSTOMERS

A final tip: the more you can make your emails feel like a personal connection with your subscribers, the better. I mentioned earlier that some of your readers will hit the "Reply" button because they don't differentiate between personal and mass emails. That's actually a good sign, and it's the type of response you should be aiming for. Here's why: as a culture that is constantly bombarded with mass advertising messages, we have a natural tendency to zone out. This is why we forward through the commercials on our DVRs, or why we drive past a billboard every day without actually noticing what's on it. You don't want your emails to fall off your readers' radar because they sound too much like every other advertisement they're being subjected to. To avoid these natural filters, you need a different, more personal approach. It's not about being overly casual; the tone and language of your emails should fit the nature of your business. Rather, it's about making readers feel like you crafted that email for their personal benefit, like they could literally reach you by hitting that "Reply" button. It's a skill that comes with practice, but if you'll keep this in mind as you write your emails and develop your marketing strategy, you'll be well equipped to make your readers feel connected—and that, in turn, leads to more customers.

EPILOGUE

The Takeaways

We've covered quite a bit of ground in the past thirteen chapters. We've gone over the inherent pitfalls of the Internet and website development. We've outlined how you can protect your company's interests in an industry that is rapidly growing and changing. We've talked at length about how to determine if a particular digital agency is one you can trust with your project and called out some warning signs to look for. We've also discussed how to get the best results once you've found an agency you can work with.

I strongly encourage you to keep this book with you and refer to it time and again as you work your way through the sometimes arduous process of selecting and working with a digital agency. For now, however, let's review four of the biggest takeaways you'll want to remember.

TAKEAWAY 1: KNOWLEDGE IS POWER

You obviously don't need to know how to write code to have an excellent working website (otherwise, you wouldn't even need this book). However, the more you understand what to look for in a good digital agency—such as knowing the right questions to ask about their workflow and process—the better equipped you will be to produce a website that will serve your business well. The people who are most vulnerable to being underserved or even swindled are those who walk into the process blindly and entrust their assets to a company who might say they know what they're doing but really don't. The early chapters in this book have hopefully armed you with the information you need during the critical stage of vetting a digital agency.

TAKEAWAY 2: YOUR CONTRACT IS YOUR PROTECTOR

The tedious details of your contract's fine print can protect you against everything from shoddy work by a digital agency to unrealistic expectations on your part. *Always* know what your contract says and doesn't say before you sign it. If the contract is missing something important, make sure to address it before moving forward. If you encounter a dispute with your agency along the way, refer back to the contract to understand your rights and the company's responsibilities. Your contract is your best defense. Don't take it lightly.

TAKEAWAY 3: COMMUNICATION IS KEY

Once you've vetted a digital agency and decided to work with them, the key to a successful working relationship with that agency is open, honest communication through clearly defined channels. The vast majority of website projects fall apart, not because of dishonest dealings, but because of miscommunication and misunderstandings. Most communication breakdowns occur because (a) expectations were not properly spelled out to begin with, or (b) the project's decision-makers aren't communicating well with each other. This is why I recommend asking lots of questions in the beginning, naming a single point of contact within your company to communicate with the agency, and agreeing on specific times and methods for regular interaction.

TAKEAWAY 4: COOPERATION IS CRITICAL

Building a website is not an automatic process; it requires you to be an active participant with the digital agency throughout the project. Remember that there are key milestones in your project that the digital agency can't reach until you have provided them the resources and information they need. Take note of your own assignments and make sure they are completed. Don't just attempt to hold the agency to their deadlines without holding yourself to your own. If you can't make a deadline, communicate this with the agency and set a new one. Flexibility is optional; cooperation and communication are essential.

CONQUERING THE WILD WEST

The Internet may still be the Wild West of our time, but that doesn't mean you have to be vulnerable. The Wild West was ultimately tamed by people who brought their own sense of order. The same can apply in the case of the Internet. Using the information and resources we've provided here, you'll be able to navigate your way to a website destined to enhance and multiply your business prospects. Best of luck!

APPENDIX

Checklist: Questions to Ask Your Digital Agency

Back in Chapter 3, we talked about the importance of asking lots of questions when vetting a digital agency to build and host your website. The checklist below is for your convenience to remind you of some of the key questions to ask, along with other questions we didn't specifically address in the book. Not all these questions will apply to every situation, but my hope is that this list will give you a track to run on when considering whether to hire an agency. Feel free to make a copy of these pages and keep them with you for future reference.

Project Questions

- Have you done any work for companies similar to mine, and may I see it?
- Will you be doing all the work in-house? If not, will you be contracting it domestically or abroad?
- What is the estimated time frame for my project?

- What are the deliverables in this website project?
- Who will I be working with on this project? (e.g., an account executive, a project manager, designers)
- What are the specific milestones for this project? What happens if one of us misses a milestone?
- Is copywriting included, or am I responsible for providing my own copy? If I'm writing the copy, when is it due?
- Will my website be search engine optimized? Who is responsible for SEO, meta tags, etc.? Who submits the site to search engines?
- When is the project considered complete by the agency?
- When do change orders occur? How do you quote change orders?
- Do you offer maintenance agreements? (i.e., how will you support me once the site is launched?)
- Is there a warranty with the website? How long does the warranty last, and how much does it cost?
- What kind of training is involved? Will you train me to manage my website, or will I need to hire someone?
- What analytics are installed with the website? Will I have Google Analytics or some other software that monitors my site? Or will I have to find that on my own?
- Will my website be available in other languages? (You should consider additional languages if more than 10 to 20 percent of your target audience speaks a language other than English.)
- Is there a pause clause? What can trigger a pause?

Payment Questions

- What are your terms of payment?
- Are payment installments triggered by calendar dates or by completion milestones?
- Is there a fixed price for this project? If not, what could make the price increase or decrease?
- What is your cancellation policy?

Technical Questions

- When my website goes down, who do I contact, and how long before I receive a response?
- Where will my website be hosted?
- If the website gets hacked, who is responsible for restoring it? Where are the backup files stored? How long are the backup files stored?
- Who is responsible for keeping my website software (e.g., WordPress, Joomla!, Drupal) up-to-date?
- If an update or upgrade breaks some sort of functionality on my website, who fixes it, and what will it cost?
- How much downtime is allowed by the Service Level Agreement? Do I get any credit for downtime?
- Do you carry insurance to cover a client's loss of income due to data or programming issues?
- What happens to my website if your company is sold or goes out of business? Where are my website assets stored, and will I be able to access them?

GLOSSARY

A/B testing.
A method of comparing two versions of a piece of digital content (e.g., newsletter, landing page, social media copy) to determine which one performs better.

Agile project management.
A project-management methodology that prioritizes scope flexibility and customer feedback throughout the process.

Asset gathering.
The collection of all materials that informs the content of a website, such as logos, stock photos, affiliate logos, mission statements, team bios, social media links, etc.

Back end.
The behind-the-scenes servers, applications, and databases that support the visible, user-facing side of a website.

Backlink.
A hyperlink that connects one website to another website. A large number of high-quality backlinks directed to a website can positively affect search engine rankings, since it can indicate that others are helping to vouch for its content.

Banner ad.
> A type of graphical display ad found on digital channels such as webpages, emails, and apps. They typically consist of an image or a multimedia object and may be either static or animated. See also *display ad*.

Bounce.
> To return an email to the sender with the notification that it was not sent. This typically happens if the recipient's inbox is full or if the recipient's email address no longer exists.

Bug.
> A coding error or flaw that causes unintended behavior in software or hardware.

Cancellation policy.
> A mutually agreed-upon contingency plan for when a business relationship must end.

Change order.
> Work that is added or removed from a contract's original scope of work.

Clickthrough.
> Each click on a link.

Clickthrough rate (CTR).
> The ratio of users that clicked on a specific link within a page, an email, or an ad to users that merely viewed the page, email, or ad.

Content.
> The text, images, sounds, and videos that make up the user experience of a website, email, or ad.

Content management system (CMS).
The application used to create, edit, and manage digital content on a website.

Conversion.
The moment a consumer performs a desired action as a result of a marketing message. The action can take many forms, such as subscribing to a newsletter, viewing a video, filling out a contact form, or buying a product.

Conversion funnel.
The journey an ecommerce consumer takes before conversion. The conversion funnel typically manifests in four steps: awareness, interest, desire, and action.

Conversion rate.
The ratio of consumers who performed a desired action as a result of a marketing message to those who did not.

Copywriting.
The process of writing meaningful marketing messages to promote brand awareness and encourage readers to take a desired action.

Cost per acquisition (CPA).
The cost to the advertiser to obtain each new customer.

Crowdsourcing.
A strategy for obtaining ideas and services through the contributions of many people, typically through an online community.

Deliverables.
The tangible or intangible goods or services provided to a customer as the result of a project.

Digital agency.
An umbrella term referring to third-party individuals or companies who play a role in setting up and/or managing a client's web development and digital marketing efforts.

Digital marketing.
The marketing of services and products to consumers using data-driven techniques within digital channels.

Discovery.
The preproduction phase of the web-development process that lays the foundation of a project by clarifying goals, gathering assets, and establishing a project timeline.

Display ad.
Visual content on a website that serves as advertising to digital consumers. There are many types of display ads, including banner ads, overlay ads, video ads, or sponsored content.

Downtime.
The percentage of time during which a website, an app, or other software is not operational.

Ecommerce.
The buying and selling of goods and services over the Internet.

Email marketing.
A type of digital marketing in which email is used to promote brands, products, and services.

Facebook Ads.
A social network advertising service by Facebook that allows the creation of ads specifically targeted to users of certain demographics.

Front end.
The visible, user-facing side of a website or app that is developed using HTML, CSS, and/or Javascript.

Funnel.
See *conversion funnel*.

Geotargeting.
A method for delivering relevant online content or display ads to users based on their location.

Google AdWords.
An online advertising service by Google that allows advertisers to pay for exposure on their search engine and partner networks.

Google Analytics.
A web analytics service by Google that lets users track, report, and analyze website traffic.

Google Images.
A search engine service by Google that enables users to search the web for images with desired keywords and perform reverse-image searches.

Intellectual property.
>Original creative output, such as art, writing, inventions, and more, that is protected from infringement by copyright, patents, trademarks, and other rights.

Keyword.
>A word or short phrase found in a website's content that allows people to find it when searching for the same keyword in a search engine.

Kill fee.
>A payment agreement within a cancellation policy that ensures contractors are paid for the time and work completed up to the point the client cancels a project.

Link building.
>A search engine optimization tactic for strengthening rankings by establishing backlinks from one site to another.

Longtail keywords.
>A keyword phrase, typically containing at least three words, that is used to target niche demographics interested in highly specific topics.

Metadata.
>The underlying information about a piece of digital content. The inclusion of metadata in webpages and images is critical for ranking well in search engines and ensuring that content is accessible to individuals protected under the American Disabilities Act.

Milestone.
>A checkpoint designated within a project timeline that is used as a signpost to evaluate progress. It may indicate a

project start or end date, internal and external reviews, submission of deliverables, and more.

Open rate.
The ratio of message recipients who opened an email to those who did not.

Organic search results.
Listings on a search engine results page that rank based on their relevancy and authority, as opposed to paid search advertising.

Paid search results.
Listings on a search engine results page that rank as a result of paid search advertising, typically facilitated through a pay-per-click campaign. See also *pay-per-click (PPC)*.

Pause clause.
A clause in a contract that gives clients the flexibility to suspend a project temporarily, either without penalty or for a fee, in the event of unexpected circumstances on the client's side that prohibit the project from moving forward.

Pay-per-click (PPC).
A paid search model in which advertisers pay a sum of money to their ad's host website every time the ad receives a click.

Phishing.
A malicious attempt to obtain a person's sensitive information by disguising electronic communication to appear as though it is coming from a trustworthy source.

Quality Assurance (QA).
The process of ensuring a high level of quality in services or products by preventing and fixing errors at every step in the production process.

Request for Proposal (RFP).
The solicitation of a proposal or quote by a company interested in procuring a service or custom product.

Request for Quote (RFQ).
See *Request for Proposal (RFP)*.

Rush fee.
An extra payment by a customer to hasten the usual turnaround time for receiving a product or service.

Scope creep.
Changes made to a project's original scope that can increase costs or cause delays.

Service Level Agreement (SLA).
A contract that defines the terms of ongoing services provided by a digital agency.

Social media.
Websites and apps that facilitate the growth of online communities and the sharing of user-generated content and communications.

Software as a Service (SaaS).
Software, such as DIY website services, licensed to users through a paid subscription or as a "freemium" service.

Statement of Work (SOW).
A contract that defines all the details of a project's timeline and deliverables.

Stock photography.
Images that may be bought or licensed for creative or commercial use, instead of hiring a photographer.

Subcontractor.
An independent worker or business, such as a copywriter or graphic designer, hired to carry out work for a larger project taken on by a principal contractor.

Template.
A reusable design framework applied to a website based on the goals and objectives of a given page.

Terms of Service (ToS).
A set of rules that a user must follow to ensure their continued use of a service.

Theme.
A semicustomizable website design that provides a basic structure and theme for the implementation of content.

Unsubscribe rate.
The amount of people who removed themselves from your email list after receiving a message.

Uptime.
The percentage of time during which a website, an app, or other software is operational.

Vector format.
    A file format for computer-generated graphics that allows high-quality resizing and future editing of text or elements. The most common vector formats are EPS, AI, PDF, and SVG.

Waterfall project management.
    A project management methodology that relies on a rigid design process characterized by a series of cascading steps completed in a set order.

Web designer.
    A software programmer who is typically responsible for handling front-end development tasks. See also *front end*.

Web developer.
    A software programmer who is typically responsible for handling the back-end construction of a website. See also *back end*.

Web hosting.
    A digital service that provides a website's server storage space and publishing access to the Internet.

Webpage.
    A single page of content on a website.

Website.
    A cohesive collection of linked webpages that typically shares a server and domain.

Wireframe.
A simplified sketch of a website that lays out the placement of content and the overall functionality and behavior of each template.

ACKNOWLEDGMENTS

I would like to express my sincere gratitude for the many people who helped write this book. To all those who helped provide support and encouragement, I am grateful. This is your book.

I would like to thank Jeff McQuilkin for countless hours of listening and guidance.

I would like to thank Catie Leary and Julie Schwietert Collazo for your help editing my ramblings into coherent prose.

I'd like to thank the team at Girl Friday Productions, especially Emilie Sandoz-Voyer, Susan Hulett, Rachel Marek, Kamila Forson, and Karen Parkin, for making this book possible.

And I'd like to thank Jason Barefoot for his continuous support from day one.

ABOUT THE AUTHOR

Drew Barton is the founder and president of Southern Web, an award-winning digital agency specializing in web development and digital marketing solutions. Over the past two decades, he has overseen thousands of successful web projects, but has also observed countless instances of malpractice within the website development industry. He has written *The Buyer's Guide to Websites* to empower business owners with a comprehensive guide that spells out exactly what to expect from a digital agency and, most importantly, how to avoid becoming a victim. He lives in Atlanta.

Website: https://drewbarton.com